Progress in Sensory Physiology

Editors: H. Autrum, D. Ottoson, E.R. Perl, R.F. Schmidt, H. Shimazu, W.D. Willis
Editor-in-Chief: D. Ottoson

Volume 10
M. Steriade, D. Paré, B. Hu, M. Deschênes
The Visual Thalamocortical System and Its Modulation by the Brain Stem Core
1990. 27 figures. VII, 124 pages.
ISBN 3-540-51380-9

Contents: Prologue.–Basic Circuits in the Visual Thalamocortical Systems and Their Physiological Aspects.–Regulatory Systems of the Brain Stem Core, Basal Forebrain, and Hypothalamus.–Neurotransmitters.–State Dependency of Visual Thalamic and Cortical Activities–References.

Volume 11
P.J. Snow, P. Wilson
Plasticity in the Somatosensory System of Developing and Mature Mammals–The Effects of Injury to the Central and Peripheral Nervous System
1991. 151 figures. XVI, 482 pages.
ISBN 3-540-52573-4

Contents: Introductory Remarks.–Plasticity in the Peripheral Somatosensory Nervous System.–Plasticity and the Mystacial Vibrissae of Rodents.–Plasticity and the Spinal Dorsal Horn.–Plasticity and the Dorsal Column Nuclei.–Plasticity and the Somatosensory Thalamus.–Plascticity and the Somatosensory Cerebral Cortex.–Concluding Remarks.–References.

Volume 12
1991. 64 figures. V, 241 pages
ISBN 3-540-52985-3

Contents: *T.B. van Wimersma Greidanus, A.B. Grossman*: Opioid Regulation of Pituitary Function.–*G.M. Innocenti*: The Development of Projections from Cerebral Cortex.–*O.R. Anderson*: Neurocognitive Models of Information Processing and Knowledge Acquisition.–*G.A. Ojemann*: Cortical Organization of Language and Verbal Memory Based on Intraoperative Investigations.

Progress in

Sensory Physiology 13

Editors: H. Autrum D. Ottoson
E.R. Perl R.F. Schmidt H. Shimazu W.D. Willis

Editor-in-Chief: D. Ottoson

Eva Syková

Ionic and Volume Changes in the Microenvironment of Nerve and Receptor Cells

With 89 Figures

Springer-Verlag
Berlin Heidelberg New York
London Paris Tokyo
Hong Kong Barcelona
Budapest

Author

Dr. Eva Syková
Institute of Physiological Regulations
Czechoslovak Academy of Sciences
Bulovka, Pavilon 11, 180 85 Prague 8, Czechoslovakia

ISBN-13: 978-3-642-76939-9 e-ISBN-13: 978-3-642-76937-5
DOI: 10.1007/978-3-642-76937-5

LC card number: 85-647743 (ISSN 0721-9156)

Softcover reprint of the hardcover 1st edition 1992

Phototypesetting: Best-set Typesetter Ltd., Hong Kong
21/3130-5 4 3 2 1 0 – Printed on acid-free paper

Contents

. . . "La fixité du milieu intérieur"

Claude Bernard (1878)

1 Introduction

Stability of the internal environment in which neuronal elements are situated is unquestionably an important prerequisite for the effective transmission of information in the nervous system. During the past decade our knowledge on the microenvironment of nerve cells has expanded. The conception that the microenvironment of neurones comprises a fluid with a relatively simple and stable composition is no longer accepted; the microenvironment is now envisaged as a dynamic structure whose composition, shape, and volume changes, thereby significantly influencing neuronal function and the transmission of information in the nervous system.

The modern conception of the neuronal microenvironment is based on the results of research over the last 20 years. The extracellular space (ECS) is comprehended not only as a relatively stable microenvironment containing neurones and glial cells (Bernard 1878), but also as a channel for communication between them. The close proximity of the neuronal elements in the CNS and the narrowness of the intercellular spaces provides a basis not only for interaction between the elements themselves, but also between the elements and their microenvironment. Substances which can cross the cell membranes can easily find their way through the microenvironment to adjacent cellular elements. In this way the microenvironment can assure non-synaptic communication between the relevant neurones. Signalization can be coded by modulation of the chemical composition of the ECS in the vicinity of the cell membrane and does not require classic connection by axones, dendrites, and synapses. This conception also encompasses the transport of chemical substances through the intracellular environment of cellular elements, e.g., glial cells, giving the spatial relationship between nerve cells a new significance.

The roots of this relatively new conception of the cellular microenvironment as a dynamically changing structure, rather than a static neuronal milieu, lie in early studies which dealt with, for instance, cellular electrical interaction, the impedance properties of the microenvironment, the influence of calcium on the cell surface, and the role of extracellular potassium concentration ($[K^+]_e$) changes in intercellular communication (Adey 1969; Schmitt and Samson 1969; Katchalsky et al. 1974; Nicholson 1980; Syková et al. 1981). It is, however, discoveries made during the past 15 years that have strengthened the conception of the microenvironment as a dynamic structure and a communication channel. These studies dealt chiefly with ionic changes in the extraneuronal space, identified by a new method which utilized ion-selective microelectrodes (ISMs) (Hinke 1959; Walker 1971) to measure, in living tissue, dynamic changes in ion concentrations in the immediate vicinity of nerve cells.

The microenvironment of nerve cells is usually identified with the ECS. It is not, however, the relationship between these two elements which is of greatest significance, but rather the division of nervous tissue into neurones, glial cells, and the ECS, i.e., three compartments, and their complex interrelationships. This review comprises the results of studies over the last 20 years, dealing with ionic changes in the neuronal microenvironment including extracellular (pH_e) changes, and discusses their physiological and pathophysiological significance. It also resolves the question of the interrelationship between neurones, glial cells, and the microenvironment and last, but not least, it presents new findings on dynamic changes in the size of the ECS accompanying neuronal activity. These questions have been considered, for the most part, in the brain, spinal cord, and sensory organs.

The findings presented unequivocally support the most recent conception of the role of the microenvironment of nerve cells. Nowadays we regard the microenvironment as a communication and modulation channel whose ionic and chemical composition, and anatomical structure significantly influence the complex function of neurones and glial cells in the nervous tissue and sensory organs.

2 Ion-Selective Microelectrodes

Until the early 1970s, all basic information on rapid changes in and sequences of membrane ionic permeability during neuronal activity was obtained primarily by the method of voltage clamping individual cells measured in vitro (Katz 1966), or by radioisotopic or other analytical methods which provided data on the total intracellular and extracellular concentrations of different ions. It was not possible to measure the actual ionic concentrations in particular areas of the nervous system during neuronal activity. Even such basic information as the resting $[K^+]_e$ in the brain was not known, due to the existence of the blood-brain barrier and other barriers. It was not possible to study hypotheses on the regulatory role of small ions during neuronal activity until conventional glass microelectrodes filled with liquid ion exchanger were introduced into neurophysiological research (Walker 1971; Vyskočil and Kříž 1972). In this chapter I would like to present a short survey of the most recent developments and procedures in the measurement of ionic activity in the CNS and receptor organs by means of ion-selective microelectrodes (ISMs).

An ISM is a miniaturized potentiometric sensor, the main component of which is an electrochemical membrane. The first ISMs were made of ion-selective glass (Hinke 1959) but their relatively small selectivity and the large size of the active surface was to their disadvantage. Today, microelectrodes with liquid membranes are the best sensors for measuring dynamic changes in the concentration of basic, biologically important ions in vivo (Fig. 1). These electrodes are widely used in electrophysiology and their development and preparation have been described in several monographs (Syková 1976; Thomas 1978; Koryta 1980; Syková et al. 1981; Ammann 1986).

An ISM basically consists of a liquid ion exchanger which forms a membrane in the tip of a glass micropipette. The ion exchanger separates two ionic solutions of different concentrations. When an ISM is introduced into the tissue or calibration solution the potential which develops across the ion exchanger membrane depends on the activity of the given ion. We use the term "selectivity" because the membrane potential is influenced by various types of ions, but from these the electrode "selects", i.e., gives priority to, a particular type. We now have at our disposal sensors (ion exchangers) for the physiological measurement of H^+, Li^+, Na^+, K^+, Mg^{2+}, Ca^{2+}, Cl^-, and HCO_3^- ion concentrations (for a review see Simon et al. 1984; Ammann 1986; Nicholson and Rice 1988; Table 1). Other important properties of ion exchangers, apart from selectivity, are signal stability, response time, and their viability (see Ammann et al. 1987). The requirement for such properties in clinical application or in basic electrophysiological research varies. Some ISMs are available commercially, but the best way to obtain microelectrodes for experimental

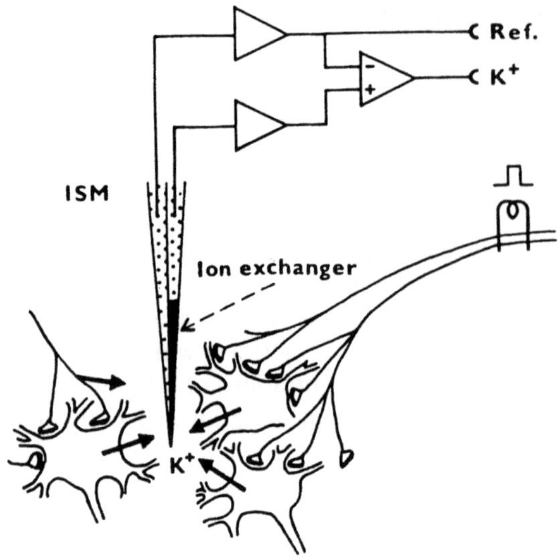

Fig. 1. Monitoring of $[K^+]_e$ changes in the CNS by means of a double-barrelled K^+-ISM (*ISM*). One channel of the ISM is filled with NaCl solution and acts as the reference electrode (*Ref.*). The tip of the other channel contains a liquid ion exchanger (Corning 477317). Electrical stimulation of afferent input produces an increase in K^+ activity in the ECS. Sources of K^+ are activated neurones and fibres. (From Syková 1986)

Table 1. Liquid membrane ion exchangers for use in microelectrodes. (From Nicholson and Rice 1988)

Ion	Exchanger	Manufacturer	Interferences	Major references
K^+	KTpClPB	Corning (477317)	Na^+ TAA^+	Numerous
K^+	Valinomycin	Fluka (60031)	Rb^+ Cs^+	Oehme and Simon (1976); Wuhrmann et al. (1979)
Na^+	ETH227	Fluka (71176)	Ca^{2+} Li^+	Steiner et al. (1979)
Ca^{2+}	ETH1001	Fluka (21048)	Mg^{2+}	Oehme et al. (1976)
Cl^-	N/A	Corning (477315)	HCO_3^- A^{-8}	Numerous
Cl^-	N/A	Corning (477913)	HCO_3^- A^-	Baumgarten (1981)
H^+	Tridodecyamine	Fluka (82500)	DNP	Schulthess et al. (1981); Ammann et al. (1981)

This table is not exhaustive. Interferences are generally more extensive than shown; only the most relevant ones are indicated.
KTpClPB, K Tetra (p-chloro)phenylborate; TAA, Tetra alkylammonium; A^-, anion.

purposes in basic neurophysiological research is to make them oneself in the laboratory.

To obtain measurements in excitable tissues, two-channel microelectrodes are used. Single-channel ISMs are unsuitable as, in addition to the potential which signals a change in the relevant ionic activity, the microelectrode also picks up biopotentials such as action potentials, field potentials, and synaptic potentials. This disturbing factor can be overcome by locating the reference electrode in the immediate vicinity of the tip of the ISM. For this reason, two-channel electrodes have proved the most satisfactory. One channel is filled with physiological saline or some other indifferent solution and acts as the reference electrode, while the other channel is the actual ISM. A brief description of the way in which most two-channel ISMs can be made is now given.

Micropipettes are prepared from prefabricated glass tubes with a diameter of 1.0–2.5 mm, which have been defatted and cleared of all impurities. We use either glass tubes with a glass septum ("theta" tubes), or pairs of tubes joined together with glass throughout their entire length. Alternatively, we join two tubes together by heating them in a flame and twisting them 360°. On a microelectrode puller, we pull out such pipettes 8–12 cm long into two microelectrodes with a tip diameter of less than 1 μm. For extracellular recording, we break off the tips, under microscopic control, to a diameter of 2–5 μm. The next essential step is to silanize the inside of the tip of the ion-selective channel so that the glass and ion exchanger form a hydrophobic bond. The microelectrode is immersed for a few seconds in silicone solution (e.g., 1%–5% tri-*n*-butyl-silane in 1-chloronaphthalene) until a silicone column of 200–1000 μm has been drawn up. To prevent silicone from entering the reference channel, its tip is filled with distilled water; prior to its immersion in silicone, a drop of water is introduced into the reference channel by means of fine polyethylene tubing, which allows the water to fill the tip by capillary action. The amount of water and length of the silicone column is controlled under a microscope. The microelectrodes are then left in an oven for 1 h at 150°–200°C. This causes the water to evaporate from the electrode and a hydrophobic silicone film is formed on the inner wall of the ion-selective channel. The siliconized microelectrodes can be stored in a dust-free environment for several weeks.

Before starting the experiment (usually on the same day or the day before), we fill the reference channel of an ISM with 0.15 mol/l NaCl solution. We then immerse the tip of the ion-selective channel in the appropriate ion exchanger – an organic substance manufactured by Corning or Fluka (see Table 1). The ion-exchanger column in the tip must be the same as, or larger than, the original silicone-coated part. When the tip has been filled with the ion exchanger, the rest of the channel is filled with the appropriate solution using a fine polyethylene tube, e.g., a K^+-selective microelectrode (K^+-ISM) is filled with 0.5 mol/l KCl solution, a Ca^{2+}-selective microelectrode (Ca^{2+}-ISM) with 0.15 mol/l $CaCl_2$ solution, and a pH-sensitive microelectrode (pH-ISM) with solution containing 40.0 mmol/l KH_2PO_4, 23.0 mmol/l NaOH, and 15.0 mmol/l NaCl at pH 7.0. Any small bubbles in the solution or between the solution and ion exchangers must be removed by means of fine hairs or by heating the air

bubble with a microcautery, which is moved towards the wall of the relevant tube under microscopic control. Before and after each experiment the microelectrodes are calibrated in solutions with different concentrations of the ions in question, usually against a background of 0.15 mol/l NaCl, i.e., in solutions of known concentrations at nearly constant ionic strength (Kříž et al. 1974).
The resistance of the ISM is generally about two orders higher than the resistance of the reference channel (10^8–10^9 ohms), and requires a special method for measuring potential changes. The ISM and the reference channels are both connected by two Ag-AgCl wires to the input of a special low noise differential amplifier with an input resistance of about 10^{14} ohms and compensation of capacitance by a negative feedback (Ujec 1988).
The use of every ISM depends on the selectivity of the electrode, and its sensitivity to the relevant ion on the one hand and to interfering ions (i.e., ions to which the ion exchanger is still sensitive) on the other (see Table 1). Ion exchangers were developed in accordance with the measurement requirements in physiological fluids, and interference between the main biologically important ions is therefore negligible. Sometimes, however, the electrodes exhibit sensitivity to transmitters, the pH, or substances introduced into the body fluids during the experiment, in which case data from the literature must be used, or the electrodes must be calibrated in solutions containing the possible interferers (Kříž and Syková 1981). The sensitivity of K^+-ISMs to substances containing a NH_4 group is used in electrophysiology, however, for measuring the size of the ECS (see Chap. 7).
In this chapter I have briefly described the ISM method, which is neither very common nor very simple, but has recently come to occupy a prominent place among those methods used to research brain function today. Intracellular and extracellular recordings utilizing ISMs have greatly enriched our knowledge of the ionic composition of biologically important ions in excitable tissues and the dynamic changes in their activity.

3 K^+ Homeostasis in the ECS

Neuronal activity is accompanied by ionic changes in the neuronal microenvironment. During action potentials or excitatory postsynaptic potentials (EPSPs), Na^+ and/or Ca^{2+} enters the intracellular compartment and K^+ leaves it. In addition to the typical Na^+ and K^+ exchange which results in concentration changes in the ECS, there are also dynamic changes in the concentration of Ca^{2+}, Cl^-, H^+, and HCO_3^-.

As a result of impulse transmission and synaptic currents $[K^+]_e$ increases under physiological conditions to between 8 and 12 mM (for a review see Syková 1983). It never exceeds 8–12 mM, i.e., the "ceiling" level, because homeostatic mechanisms effectively remove K^+ from the ECS; in this way the stability of the neuronal microenvironment is assured and K^+ buildup during neuronal activity is kept to a minimum. Since K^+ cannot be transformed to another compound, it follows that it is transported from one compartment to another. Eventually it has to be returned to the neurones from which it was released; otherwise, the neurones would be depleted of their K^+. Chapter 4 discusses the role of K^+ as a powerful modulator of impulse transmission in the CNS and receptor organs and it may well be possible that it serves as an important signal between neurones, and between neurones and glia, as already pointed out by Hertz (1965), and Kuffler, Nicholls, and Orkand (1966). On the other hand, K^+ accumulation during pathological events (see Sect. 4.3.6) can dramatically impair neuronal and glial cell function. The K^+ homeostatic regulatory mechanisms maintaining appropriate composition of the cell's microenvironment are therefore of primary importance in the CNS.

3.1 Stability of K^+ in the Extracellular Fluid

The homeostatic system of the brain and spinal cord controls the chemical composition of the fluid in the vicinity of neurones and glial cells. Compared to plasma, the chemical composition of extracellular fluid is relatively stable and its stability is assured by a dual cell population, i.e., the endothelial cells of the brain capillaries and the epithelial cells of the choroid plexus. The epithelial cells which produce the CSF are interconnected and thereby form a barrier which prevents the infiltration of certain molecules and ions (Davson 1976; Rapoport 1976).

The maintenance of a stable extracellular fluid composition in the CNS in undoubtedly very important. If it were not stable, disturbances in signalization and integration in the nervous tissue could very easily occur. Particularly important is the Na^+, K^+, Ca^{2+}, Mg^{2+}, Cl^-, H^+, and HCO_3^- ion concentration in both the CSF and extracellular fluid, where the level is carefully controlled and, for the most part, maintained independently of the plasma level of these ions (Kemény et al. 1961; Ames et al. 1964; Fencl et al. 1966; Davson 1967; Katzman and Pappius 1973).

The membrane potential of excitable cells depends on the ion concentration gradient between the internal and external environment of the cell membrane, and on the permeability of the cells membrane. Nerve cell depolarization that is caused by either action potentials or EPSPs results from an influx of Na^+ and/or Ca^{2+} into the cell. The number of ions exchanged during one such action potential is, however, so small that the effect on the ion concentration in the ECS is insignificant. With each depolarizing potential the neurone gains Na^+ and loses K^+. The equilibrium potential for K^+ (E_k) is determined by the Nernst equation:

$$E_k = \frac{RT}{zF} \log_e \frac{[K^+]_e}{[K^+]_i}$$

in which $[K^+]_e$ is the K^+ concentration outside, and $[K^+]_i$ is the K^+ concentration inside the cell membrane; R is the gas constant, T is the absolute temperature, z is the valency, and F is Faraday's constant. The equilibrium potential for K^+ plays an important role in the maintenance of the cell's resting potential, since any increase in $[K^+]_e$ can significantly affect the function of the neurones and glial cells.

The regulatory mechanisms which prevent sudden and dramatic changes in $[K^+]_e$ in the CSF and ECS are very important because of the relative frequency with which acute hypokalaemia and hyperkalaemia occur in the organism. However, neither spontaneous, nor experimentally induced acute or chronic changes in the plasma K^+ concentration are reflected in the K^+ level in the CSF or extracellular fluid (Ames et al. 1965; Bradbury and Davson 1965; Bradbury and Kleeman 1967; Husted and Reed 1976). In rats given a K-free diet, the total K^+ content of the muscles fell by 34%, whereas K^+ content of the brainstem fell by only 1.9% and that of the forebrain fell by only 4.0% (Nattie 1977).

The epithelial cells of the choroid plexus do not act purely as a barrier; they also regulate the K^+ level in the CSF by means of an active transport system – the Na^+/K^+ pump (Bradbury and Štulcová 1970; Wright 1972; Johanson et al. 1974). The stable composition of the CSF assures an optimal microenvironment for neurones and glial cells, not only in man and other mammals, but also in amphibians (Cohen et al. 1968). Table 2 shows the differences in concentration of various ions in human plasma and CSF.

The "resting" $[K^+]$ in the extracellular fluid filling the intercellular spaces does not differ significantly from values in the CSF. Using K^+-ISMs, it was demonstrated that the resting $[K^+]$ in the ECS of the brain and spinal cord of mammals varied from 2.6 to 3.8 mmol/l (Krnjević and Morris 1972; Vyskočil et al. 1972; Vyklický et al. 1972; Lux and Neher 1973; Prince et al. 1973; Futamachi et al. 1974). Variation of $[K^+]$ depends on the spontaneous activity

Table 2. Na^+, K^+, Ca^{2+}, Mg^{2+}, Cl^-, and HCO_3^- ion concentrations (mEq/kg water) in human plasma and cerebrospinal fluid. (After Davson 1976)

Ion	Concentration in plasma	Concentration in CSF
Na^+	150	147
K^+	4.63	2.86
Ca^{2+}	4.70	2.28
Mg^{2+}	1.61	2.23
Cl^-	99	113
HCO_3^-	26.8	23.3

CSF, cerebrospinal fluid.

of the nerve cells, however, as we showed in the mesencephalic redicular formation (MRF) and in the dorsal horns of the spinal cord (Syková et al. 1974, 1983; Svoboda et al. 1988; see Sect. 4.1.5).

3.2 Sources of $[K^+]_e$ Increases

$[K^+]_i$ is many times greater than $[K^+]_e$. Before ISMs were introduced into neurobiology, intraneuronal K^+ activity was determined indirectly, by calculating the total ion content and estimating relative extracellular fluid volume (Katzman and Pappius 1973). Measurement by means of double-barrelled K^+-ISMs has since shown that $[K^+]_i$ in frog motoneurones varies from 74.8 ± 2.5 mmol/l (Chmykhova 1982) to 87.8 ± 34.1 mmol/l (Bührle and Sonnhof 1981). These $[K^+]_i$ values in the brain and spinal cord of vertebrates are 25–45 times higher than the $[K^+]_e$ level. It is thus evident that just a small quantity of intracellular K^+ can significantly alter $[K^+]_e$ in narrow intercellular clefts only 15–30 nm wide.

The accumulation of potassium ions in the intercellular clefts is the outcome of neuronal activity. Stimulated or spontaneously active neurones, non-myelinated fibres, and non-myelinated endings of myelinated axones are the main sources of K^+ accumulation. Neuronal somatod endritic membranes and non-myelinated fibres release K^+ with every impulse, for instance, the total amount of K^+ released during one impulse from 1 cm^2 squid axonal membrane is 4.0 pmol/l (Keynes 1951; Keynes and Lewis 1951; Shanes 1954). Similarly, Keynes and Ritchie (1965) showed that 1.1 pmol/l K^+ was released during a single impulse from 1 cm^2 membrane of mammalian non-myelinated fibres. In experiments on squid giant nerve fibres, Frankenhauser and Hodgkin (1956) submitted the first evidence showing that small amounts of K^+ released from the fibres into the intercellular space during a single action potential do not immediately diffuse away from the vicinity of the fibre, but accumulate in the intercellular clefts around the active fibre, thereby raising $[K^+]_e$ by 1.6 mmol/l. These authors

utilized the finding that the increase in K^+ conductance and the inactivation of Na^+ conductance accompanying repolarization lead to a situation where the membrane potential of the neurone actually drops below (undershoots) the resting membrane potential and approaches the equilibrium potential for K^+. Adelman and Fitzhugh (1975) estimated that the $[K^+]_e$ value in the periaxonal space rose during a single impulse by 1.0 mmol/l.
The introduction of K^+-ISMs led to a dramatic advance in the determination of K^+ changes during neuronal activity in the CNS. The increase in K^+ during spontaneous activity of neurones in the MRF could be measured directly (Syková et al. 1974) and, in the vicinity of a spontaneously active neurone, $[K^+]_e$ was seen to rise by 0.01–0.02 mmol/l. In the dorsal horns of the spinal cord, the stimulation of a peripheral nerve or a dorsal root with a single electric pulse excites a whole population of interneurones, thereby causing $[K^+]_e$ to rise by up to as much as 0.5 mmol/l (Kříž et al. 1975). A similar increase in $[K^+]_e$ after a single electrical pulse or after adequate stimulation of afferent input was found during activity of Purkinje cells in the cerebellum (Bruggencate et al. 1976). These changes measured during a brief increase in neuronal activity are no doubt actually much greater, however. An increase in $[K^+]_e$ in a space only 15–30 nm wide cannot be measured exactly using an electrode with a tip diameter of 1–4 um, since such an electrode cannot be placed close enough to the surface of a neurone or a nerve fibre (Fig. 2).

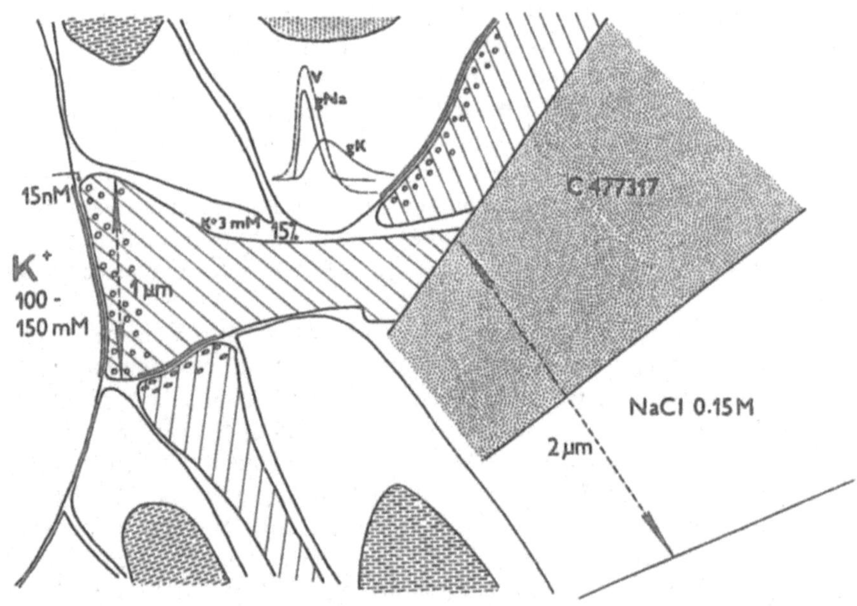

Fig. 2. Experimental situation, when measuring $[K^+]_e$ changes with a double-barrelled K^+-ISM (tip diameter 2 μm). The tip of the ion-selective channel is filled with a liquid ion exchanger (Corning 477317) and the reference channel with 0.15 mol/l NaCl solution. The synaptic cleft between the neurones, fibres, and glial cells is about 15 nm wide. $[K^+]_i$ is 100–150 mmol/l and $[K^+]_e$ is 3 mmol/l. (From Syková 1981)

Transient changes in $[K^+]_e$ are greater during repeated stimulation of cells or fibres, especially when high stimulation frequencies are employed. The increase in $[K^+]_e$ which results from the repetitive stimulation of many elements can be effectively measured with a K^+-ISM – even though the electrode has created a dead space around its tip and measurement of the true $[K^+]_e$ value is slowed down by the diffusion of K^+ ions released into this space from activated cells and fibres. During repetitive stimulation $[K^+]_e$ rises until a steady state is established, which is the result of concurrent release and clearance of K^+ (see Fig. 4). This ceiling level, or maximum attainable change in $[K^+]_e$ is probably a fairly accurate representation of the actual increase in $[K^+]_e$ (Kříž et al. 1975; Syková 1981). Experiments have shown that if we rule out anoxia, ischaemia, and epileptiform activity, stimulation leads to a $[K^+]_e$ increment of not more than 7–9 mmol/l in the CNS of adult mammals. These relatively large increases in $[K^+]_o$ were found during repetitive stimulation in the cerebral cortex, cerebellum, hippocampus, and spinal dorsal horns (see Syková 1983).

Synaptic currents, as well as impulse activity, can be a source of an increase in $[K^+]_e$. Gamma-aminobutyric acid (GABA) stimulates K^+ release from rabbit cortex (Brinley et al. 1960a), and, along with several other putative transmitters and neuropeptides, also raises $[K^+]_e$ in the spinal cord (Kudo and Fukuda 1976; Syková 1978, 1983; Vyklický et al. 1982; see Sect. 4.1.6).

3.3 Redistribution of Extracellularly Accumulated K^+

When they lose K^+, active neurones induce a change in the ionic composition of the extracellular fluid which significantly affects the function of neighbouring elements. This change in ionic composition may be necessary for the interaction and integration of neuronal activity; however, the stability of brain and spinal cord functions requires the prompt and exact renewal of the composition of the neuronal environment. There are, in fact, two problems which must be resolved simultaneously: (1) excessive and prolonged changes in $[K^+]_e$ must be prevented to avoid disturbances of function and irreversible changes, and (2) neurones must regain the K^+ released during their activity, or they will no longer be able to function. Neurones are, in the event, very good at regulating their internal ionic composition. Any decrease in $[K^+]_i$ and increase in $[Na^+]_i$ is quickly compensated for by means of active transport mechanisms. Fluctuations in $[K^+]_e$ are also compensated for, by either (a) active transport across the cell membranes, (b) the voltage gradient in the tissue, (c) diffusion in the ECS, (d) transport across the blood-brain barrier, or (e) water shifts between the various tissue compartments (Fig. 3).

3.3.1 Role of Active Transport

The intracellular K^+ released by active neurones must eventually be returned to the cell. The main mechanism by which the resting intra- and extracellular

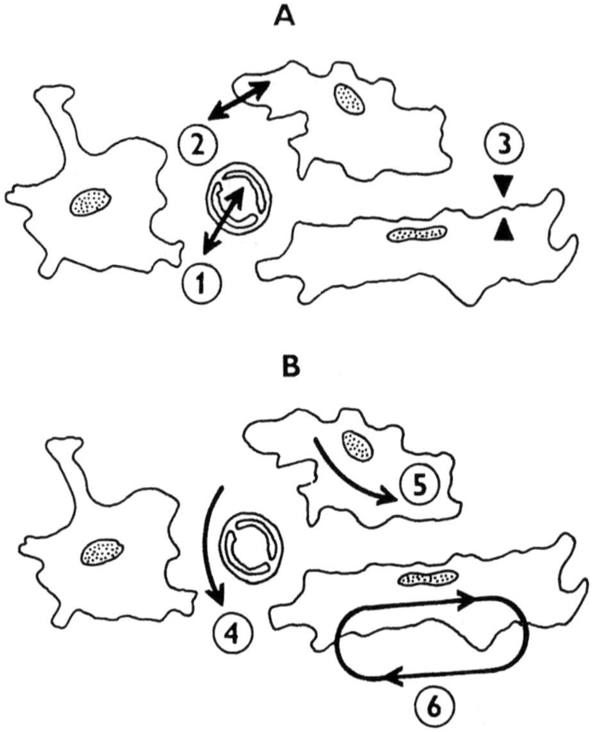

Fig. 3A,B. Mechanisms influencing $[K^+]_e$. **A** Mechanisms involving various compartments, but independent of the K^+ gradient between different tissue areas. **B** Mechanisms dependent on the K^+ gradient: *1*, K^+ exchange across the capillary wall; *2*, K^+ exchange between the intracellular and extracellular compartments, including active transport; *3*, concentration changes caused by water shifts across the cell membranes; *4*, diffusion in the ECS; *5*, diffusion in the intracellular space; *6*, K^+ transport associated with a current flow. (After Gardner-Medwin, in Nicholson 1980)

ionic composition is restored is the Na^+/K^+ pump. An increase in $[Na^+]_i$ and an increase in $[K^+]_e$ both stimulate active transport in excited neurones – Na^+ is pumped out of the cells and K^+ is pumped into them. Glynn and Karlish (1975) showed that the activity of the pump is proportional to $[Na^+]_i$ if $[K^+]_e$ was constant. Active transport may, however, be stimulated by an increase in $[K^+]_e$ alone, indicating that active transport can be stimulated in passive, as well as in active neurones.

On the basis of a series of measurements carried out using ISMs, it can be concluded that active transport is one of the main mechanisms by which K^+ ions which have accumulated in the ECS are transported back to the neurones. Repetitive stimulation in the CNS leads to an increase in $[K^+]_e$ of several millimoles per litre (see Sect. 4; Syková 1981, 1983), but in the mammalian brain and spinal cord, discontinuation of stimulation is followed by a drop in $[K^+]_e$ to below the original resting value. This gives rise to a $[K^+]_e$ "undershoot" (Fig. 4A; Heinemann and Lux 1975; Kříž et al. 1975; Krnjević and

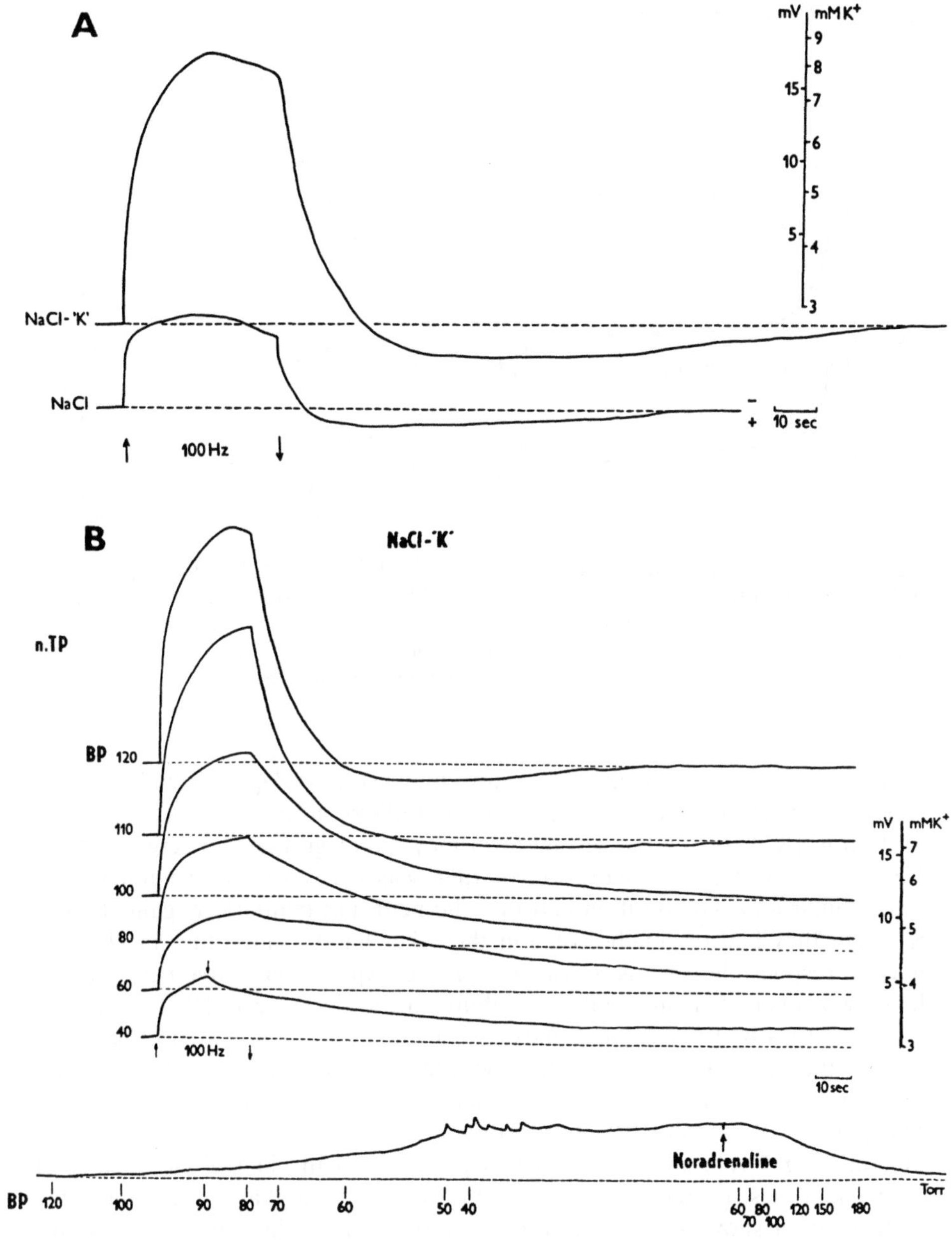

Fig. 4A,B. Typical $[K^+]_e$ changes in the cat spinal cord during peripheral nerve stimulation. **A** The increase in $[K^+]_e$ (*NaCl–"K"*) during 30 s electrical stimulation at 100 Hz. The extracellular potential change was measured simultaneously by means of the reference channel of the double-barrelled K^+-ISM (*NaCl*). **B** The effect of hypoxia on the $[K^+]_e$ increase produced by stimulation of the posterior tibial nerve (*n. TP*) at 100 Hz, recorded during a decrease in blood pressure (*BP*) from 120 to 40 mm Hg. The tibial nerve was stimulated for a total of 28 s; at a BP of 40 torr for only 15 s. Note that, at a BP of 60 and 40 mm Hg, the resting $[K^+]_e$ level rises after stimulation. Bottom trace, increase in the resting $[K^+]_e$ during hypoxia and its return to the initial level after injecting noradrenaline. (From Kříž et al. 1975)

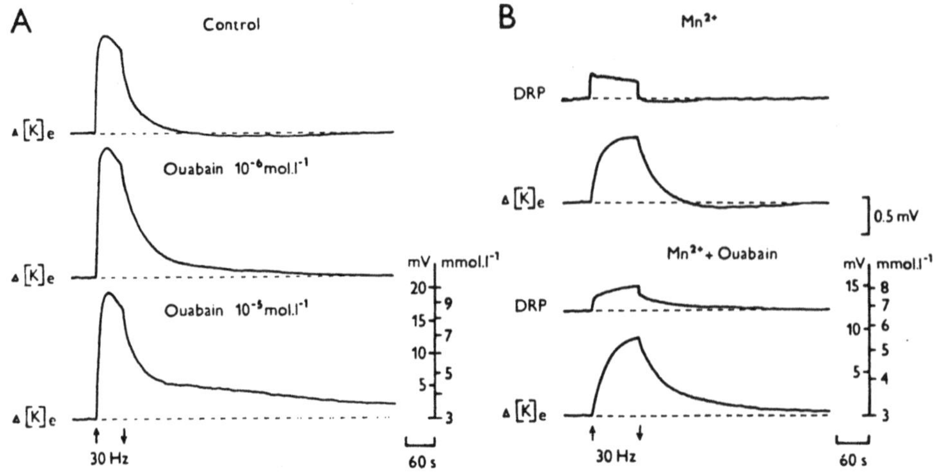

Fig. 5A,B. The effect of ouabain on the increase in $[K^+]_e$ and on dorsal root potentials (*DRP*) evoked in the dorsal horns of the isolated frog spinal cord by electrical stimulation at 30 Hz. **A** $[K^+]_e$ before (*control*) and 30 min after administering ouabain in 10^{-6} and 10^{-5} mol/l concentration. **B** The effect of ouabain ($5 \cdot 10^{-5}$ mol/l) on $[K^+]_e$ and DRP in the presence of 2 mmol/l Mn^{2+}. (From Syková et al. 1985a)

Morris 1975). This transient drop below the resting $[K^+]_e$ clearly shows that active transport processes participate in the redistribution of K^+.

A drop in $[K^+]_e$ below the resting level may be a sign that the activity of the Na^+/K^+ pump was stimulated longer than was necessary for the renewal of intracellular ionic composition and that the neurones temporarily gained more K^+ than they had originally lost. On the other hand, some other mechanism may be responsible for this transient decrease. Apart from active neurones, at least two other elements could contribute to the decrease in $[K^+]_e$: (a) non-activated (passive) nearby neurones and (b) adjacent glial cells. In the spinal cord of the cat, however, undershoot was found only in the immediate vicinity of active neurones, i.e., in the deeper layers of the spinal dorsal horns. Undershoot is not directly proportional to the increase in $[K^+]_e$, i.e., the amount of K^+ released, and it occurs primarily in the immediate vicinity of active neurones. Furthermore, there is no evidence that glial cells are unevenly distributed or show specialization in different layers of the spinal cord, Active neurones are thus probably largely responsible for the reuptake of K^+ ions which have been released into the ECS during stimulation. This hypothesis is further supported by the finding that undershoot is blocked during partial anoxia or a drop in blood pressure (Fig. 4B). Last, but not least, undershoot is also blocked after the administration of ouabain, a selective inhibitor of Na^+/K^+ ATPase (Fig. 5; Kříž et al. 1975; Krnjević and Morris 1975; Syková et al. 1985a). Any process or substance which blocks the activity of the Na^+/K^+ pump lengthens the time taken by $[K^+]_e$ to return to the resting level and thereby prevents undershoot. This applies, for example, to a decrease in body temperature and to the administration of barbiturates and naloxone (Lewis and Schuette 1975a; Syková et al. 1976, 1985a; Morris 1978; Sick and Kreisman 1981; Hájek et al. 1985). In mammals, oxidative metabolism exhibits the same time course as K^+ accumulation in the ECS of the spinal cord, the cerebral cortex, and the hippocampus; this indicates that the greater part of increased metabolic activity is related to the restoration of ion gradients (Rosenthal and Somjen 1973; Mayevsky et al. 1974; Lewis and Schuette 1976; Somjen et al. 1976; Rosenthal et al. 1979).

It can be concluded that active reabsorption of K^+ can be stimulated by an increase in both $[Na^+]_i$ (active elements) and $[K^+]_e$ (passive elements). Although the Na^+/K^+ pump is an important transmembrane transport mechanism for the clearance of K^+ which has accumulated in the ECS, some ions may escape active reabsorption by following the gradient in the intracellular space or by being transported by glial cells. Na^+/K^+ exchange across the cell membrane may, alternatively, be the result of purely passive fluxes (Nicholson 1980). Non-activated cells, especially the glial cells via the "spatial buffer mechanism", also contribute passively to K^+ clearance and to Na^+/K^+ exchange.

3.3.2 *Role of Glial Cells in K^+ Homeostasis*

The glial cells in the CNS play an important role in K^+ homeostasis by means of both passive and active mechanisms (for review see Nicholson 1980; Treherne 1981; Syková 1983; Walz 1989). Passive processes comprise K^+ movements towards Donnan's equilibrium, and the "space-dependent buffer mechanism" which represents the movement of K^+ along an electrochemical gradient. K^+ ions enter the glial cells in places where $[K^+]_e$ is raised and leave them in places where it is low (Orkand et al. 1966; Coles and Orkand 1983; Gardner-Medwin 1983a,b).

3.3.2.1 *The Spatial Buffer Mechanism*

Orkand et al. (1966) postulated that if an increase in $[K^+]_e$ appeared in the ECS of nervous tissue, depolarization of the glial cells could cause a current to flow through them. This current flux, carried by K^+, helps to redistribute K^+ ions from parts of the ECS where their concentration is raised to places where it is low. In this way, K^+ can be transported by the cytoplasm of glial cells from regions where it has accumulated during stimulation (i.e., the vicinity of active neurones) to regions where its concentration is unaltered and therefore lower (i.e., the neighbourhood of non-active neurones). This K^+ redistribution mechanism, known as the spatial buffer mechanism, is a purely passive process; it has no energy requirements and will function as long as there is a K^+ gradient in the tissue. This current concept is based on three basic properties of the glial cell membrane: (a) glial cell membranes are highly permeable for K^+; (b) the current across glial cell membranes is carried exclusively by K^+; and (c)

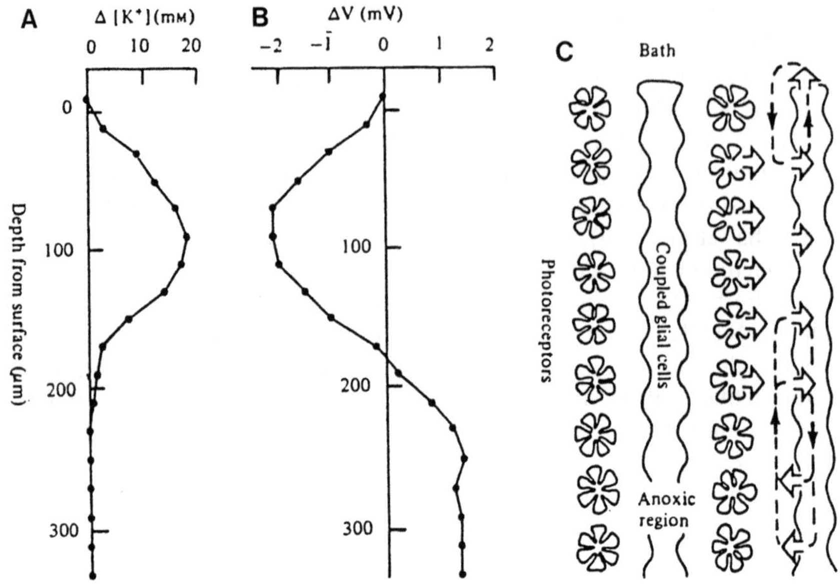

Fig. 6A–C. Spatial buffering currents generated by light-evoked $[K^+]_e$ increases in the cut drone retina. Measurements with K^+-ISMs demonstrate that light stimulation generates an increase in $[K^+]_e$ (**A**) and negative field potential (**B**) within the drone retina. Both responses reach maximal values at approximately 100 μm below the cut surface of the preparation. A schematic representation of the drone eye (**C**), drawn to the same vertical scale as **A** and **B**, illustrates the origin of the field potential. K^+ released from photoreceptors (*rosettes*) enters glial cells in the region of maximal $[K^+]_e$ increase. K^+ current exits from distant regions of the glial syncytium. The two loops of current established by these K^+ fluxes (*broken lines*) generate the extracellular field potential measured in **B**. (From Gardner-Medwin et al. 1981)

glial cells are electrically coupled. Between individual glial cells there are low-resistance connections (gap junctions) which allow K^+ to pass with ease through the glial syncytium (Cohen 1970; Kuffler and Nicholls 1966; Syková and Orkand 1980; Gutnick et al. 1981; Orkand et al. 1981; Coles and Orkand 1983; Kettenmann et al. 1983a). The elongated shape of the glial cells and their interconnection in a syncytium makes them ideal for spatial buffering. The glial cell syncytium tends to be isopotential. The local accumulation of K^+ described in the spinal cord and brain of vertebrates affects only a part of the glial cell membrane; the rest of the syncytium is surrounded by a low $[K^+]_e$. In that part of the glial cell syncytium where depolarization is greater during the increase in $[K^+]_e$, K^+ – as a major current carrier – enters the glia and current loops are formed through the cell cytoplasm, the gap junctions, and the ECS. Any ion will carry the current in the cytoplasm and across the gap junctions, but K^+ is the major carrier ion, while in the ECS the current is carried mainly by Na^+ and Cl^- (Fig. 6).

Diffusion experiments in rat cerebellum clearly demonstrate that glial cells are capable of spatial buffering (Nicholson and Phillips 1981). The spatial buffer

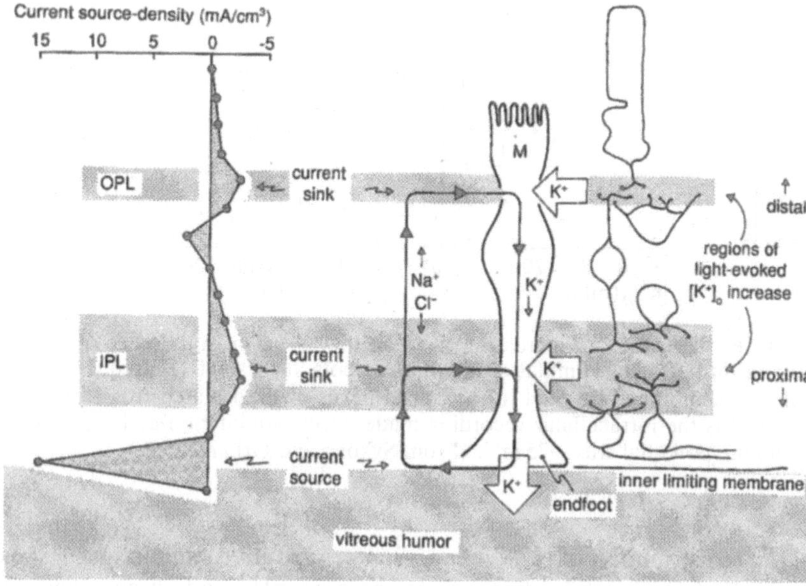

Fig. 7. Generation of ERG b wave by Müller cells. The result of a current source-density analysis of the b wave (*left*) demonstrates that ERG response is generated by two current sinks in the inner and outer plexiform layers and by a current source at the vitreal surface of the retina. The current sinks arise when K^+ current enters the cell in regions of light-evoked increase in $[K^+]_e$ (*open arrows at right*). Almost all of this current exits from the endfoot of the Müller cell. This efflux is recorded as a current source. The current flow through ECS established by these K^+ fluxes (*solid lines*) generates the b wave. This pattern of current flow also leads to the transfer of excess K^+ from the plexiform layers to the vitreous humor. *M*, Müller cell; *OPL*, outer plexiform layer; *IPL*, inner plexiform layer. (From Newman 1985a)

mechanism has also been studied in the retina (Gardner-Medwin et al. 1981; Coles and Orkand 1983; Newman 1986; Karwoski et al. 1989). Newman and his colleagues demonstrated that the Müller cells of the salamander retina were capable of spatial buffering by what Newman termed "K^+ siphoning" (Fig. 7). Measurements made with ISMs in the spinal cord have shown that the glial cells of vertebrates are depolarized in vivo when (a) $[K^+]_e$ is raised by adding K^+ to the perfusion medium, or (b) K^+ is released into the ECS as a result of the repetitive stimulation of afferent input (Fig. 8; Syková and Orkand 1980). The presence of voltage-dependent potassium channels has been demonstrated in oligodendrocyte and astrocyte tissue cultures (Kettenmann et al. 1982; Sonnhof and Schachner 1986), and there is also evidence that astrocytes have an extremely high K^+ conductance in their endfeet (Newman 1986).

Experiments carried out in the spinal cord show that the glial spatial buffer mechanism probably applies here as well (Syková et al. 1988a). In the isolated frog spinal cord, we studied the effect of an increase in pCO_2 on (a) the time course of changes in $[K^+]_e$ after electrical stimulation of the dorsal roots; and

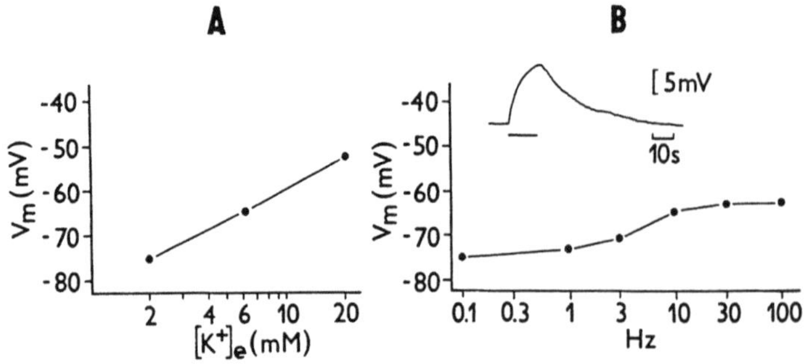

Fig. 8A,B. The effect of increasing $[K^+]_e$ in the perfusion solution to 6 and 20 mmol/l (**A**) and the effect of stimulating the dorsal roots at 0.1–100 Hz frequency (**B**) on the glial cell membrane potential in the dorsal horns of the isolated frog spinal cord. The *inset* in **B** is the intracellular recording made during stimulation at 100 Hz. The resting membrane potential was −75 mV. (From Syková and Orkand 1980)

(b) the redistribution of $[K^+]_e$ changes after repetitive stimulation. Exposure to a raised pCO_2 level leads to a decrease in pH; and disconnects coupling between the cells of various tissues, including the neuroglia (Turin and Warner 1980; Spray et al. 1981; Tang et al., 1985). Figure 9 shows that $[K^+]_e$ changes in the isolated frog spinal cord are affected when the spatial buffer mechanism is disturbed by an increase in pCO_2. A tenfold increase in pCO_2 in the Ringer solution perfusing the spinal cord resulted in a slower clearance of the K^+ accumulated by stimulation in the ECS, and in a narrowing of the area to which K^+ spread from the region of the stimulated neurones and fibres (Fig. 10). Furthermore, it has been shown in recent studies that stimulation of the afferent input produces much greater changes in $[K^+]_e$ in the immature rat spinal cord than in the adult spinal cord (Walton and Chesler 1988; Jendelová and Syková; 1991 see Fig. 76). Greater stimulus-evoked $[K^+]_e$ changes have also been found in the immature rat optic nerve (Connors et al. 1982; Ransom et al. 1985a,b) and in the cerebral cortex (Mutani et al. 1974; Hablitz and Heinemann 1987, 1989). The smaller $[K^+]_e$ transients in adult rats could be the result of glial cell participation in K^+ homeostasis; the number of astrocytes and oligodendrocytes changes, and their maturation and myelinization of fibres occur postnatally (Gilmore et al. 1982; Connors et al. 1982; Sims et al. 1985; Yamate and Ransom 1985). These results indicate that the effective clearance of K^+ from the ECS of the dorsal spinal horns after either electrical or adequate stimulation, or in pathological states, includes K^+ influx through glial cells along a concentration gradient, i.e., the glial cell spatial buffer mechanism. This leaves us with the question of the relationship between glial cell clearance of accumulated K^+ and its removal by the neuronal Na^+/K^+ pump. I conclude that the spatial buffer mechanism is evidently much faster, and hence more effective, than mere diffusion in the ECS (Nicholson 1980; Gardner-Medwin 1983a,b). There is practically no lag in origination of the current carried by K^+, and about 80% of the K^+ moving through the tissue is

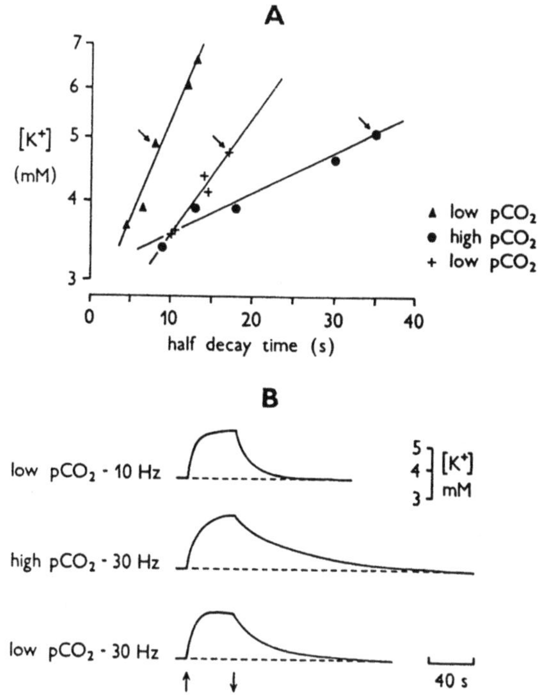

Fig. 9A,B. The effect of increasing CO_2 (50% CO_2/50% O_2) in the Ringer solution on the stimulation-induced increase in $[K^+]_e$ and on its redistribution in the dorsal horns of the isolated frog spinal cord. **A** The $[K^+]_e$ increment was induced by tetanic electrical dorsal root stimulation (40 s) at different frequencies. The recovery half-time of stimulation-induced $[K^+]_e$ changes was measured at a depth of 400 μm below the dorsal surface of the spinal cord. *Triangles*, recordings in normal pCO_2 (5%) during stimulation with frequencies of 3, 10, 30, 100, and (again) 3 Hz; *circles*, recordings 10–50 min after raising pCO_2 (50%) during stimulation at frequencies 3, 10, 30, 100, and (again) 10 Hz; *crosses*, partial recovery during 10–50 min in normal pCO_2 (5%) during stimulation at 3, again 3, and 10, 30, and 100 Hz. **B** Examples of $[K^+]_e$ changes, which were approximately the same size (2 mmol/l) in a low and a high pCO_2, at stimulation frequencies of 10 and 30 Hz. The recovery half-times measured from these curves are denoted in **A** by *arrows*. (From Syková et al. 1988a)

transported by glial cells, only the remaining 20% going through intercellular clefts (Gardner-Medwin et al. 1979; Gardner-Medwin 1983a,b). The spatial buffer mechanism is therefore an important mechanism by which the glia controls K^+ homeostasis in the CNS and in the receptor organs.

3.3.2.2 Active K^+ Transport

Recently, there have been a renewed studies into the question of whether or not glial cells remove K^+ from the ECS by active transport – an idea which is actually older than the hypothesis which describes the passive role of glia in K^+

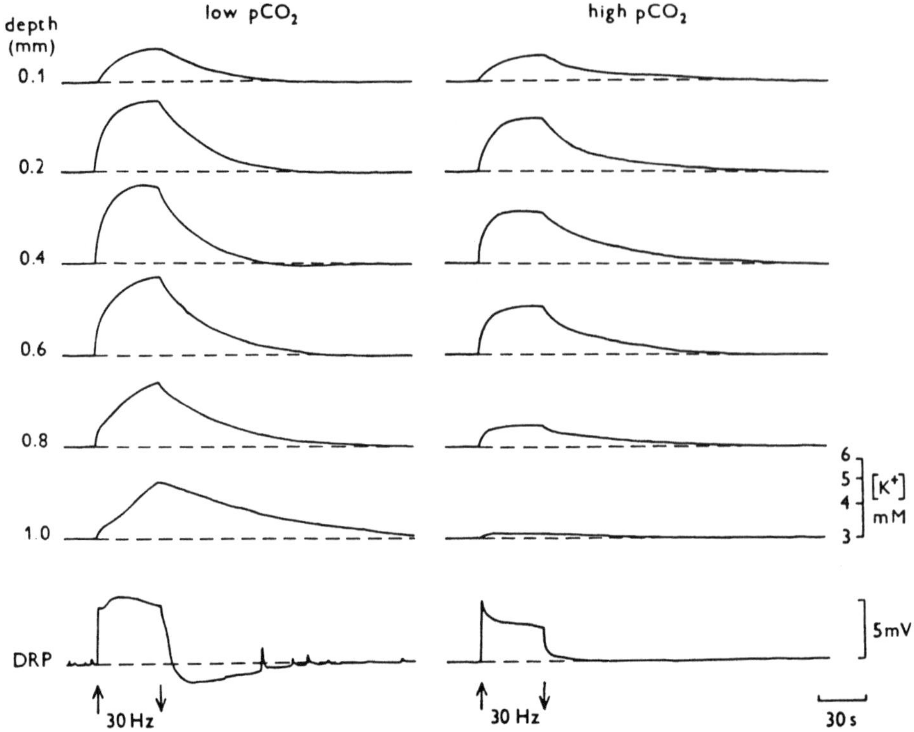

Fig. 10. The effect of a pCO_2 increment from 5% to 50% on the spatial distribution of $[K^+]_e$ changes in the isolated frog spinal cord and on dorsal root potentials (*DRP*) evoked by tetanic dorsal root stimulation (30 Hz, 40 s). An increase in pCO_2 inhibits stimulation-induced changes, especially at a depth of 0.8 and 1.0 mm, i.e., at a depth corresponding to the superficial layers of the ventral horns. It should be mentioned that the slow component is missing from the DRP recording after raising pCO_2. (From Syková et al. 1988)

clearance (Cummins and Hydén 1962; Hertz 1965; Henn et al. 1972; Hertz and Schousboe 1975; Coles 1981; Walz and Hertz 1982; Kettenmann et al. 1983b, 1987). Tang et al. (1980) stated that the addition of K^+ to glial cells whose intracellular K^+ concentration had been artificially reduced activated the electrogenic pump, which, as in neurones, is stimulated by an increase in $[Na^+]_i$ and required extracellular K^+. Waltz and Hertz (1982) demonstrated ouabain-sensitive and ouabain-resistant K^+ uptake in astrocyte tissue cultures. In immunologically identified oligodendrocytes from the spinal cord of embryonic mice, it was shown that K^+ uptake by glial cells is indeed, at least partly, assured by active transport through the exchange of K^+ for Na^+ (Kettenmann et al. 1987). Na^+/K^+ ATPase activity is stimulated by either an increase in $[Na^+]_i$ or a decrease in $[K^+]_i$. When Na^+ was introduced iontophoretically into glial cells, there was a decrease in $[K^+]_i$. (Fig. 11). The reuptake of K^+ was, however, blocked by ouabain, suggesting that the mech-

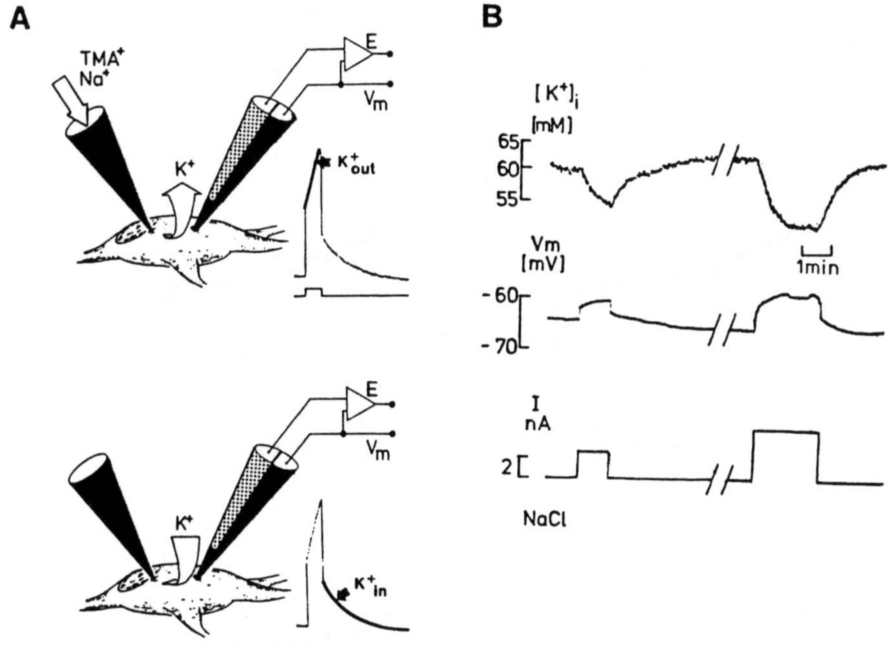

Fig. 11A,B. Iontophoretic depletion of $[K^+]_i$. **A** Experimental set-up. Two microelectrodes were introduced into an oligodendrocyte in a mouse embryonic spinal cord tissue culture – one for the application of a current (*left*) and the other for recording the membrane potential (V_m) or $[K^+]_i$ (E). During the injection of TMA^+ or Na^+, K^+ is drawn out of the cell. As the membrane potential is recorded, progressive depolarization can be seen during application of the current (K^+_{out}). After application of the current the membrane is depolarized compared with the original membrane potential, but subsequently it becomes repolarized (K^+_{in}). **B** Recordings of the membrane potential (V_m) and $[K^+]_i$ during injection of the depolarizing current (I) with an electrode filled with NaCl. The membrane is depolarized, $[K^+]_i$ falls, and when the current is discontinued it returns to the original level. (From Kettenmann et al. 1987)

anism by which it is assured is Na^+/K^+ ATPase dependent (Fig. 12A). After injecting Na^+ into the cells, K^+ reuptake was not blocked in a chloride-free medium (Fig. 12B), even in the presence of the Cl^- transport blockers furosemide, SITS, and DIDS (Kettenmann et al. 1987). In astrocytes, however, Kimelberg et al. (1979a,b) showed that K^+ was transported together with Cl^-; this was also demonstrated in the glial cells of the bee's retina (Coles and Orkand 1984). Walz and Hinks (1986) found experimental evidence for a transmembrane Na^+ cycle in cultured astrocytes, which is shown in Fig. 13. The proposed cycle involves the Na^+/K^+ pump, which is stimulated by an increase in $[Na^+]_i$ and $[K^+]_e$. Na^+ is pumped out of and K^+ is pumped into the glia, while concomitantly NaCl and KCl carriers transport these ions inwards. This mechanism may lead to KCl accumulation and the uptake of water to compensate for the increase in osmolarity, with resultant swelling of the glial cells (see Chap. 7).

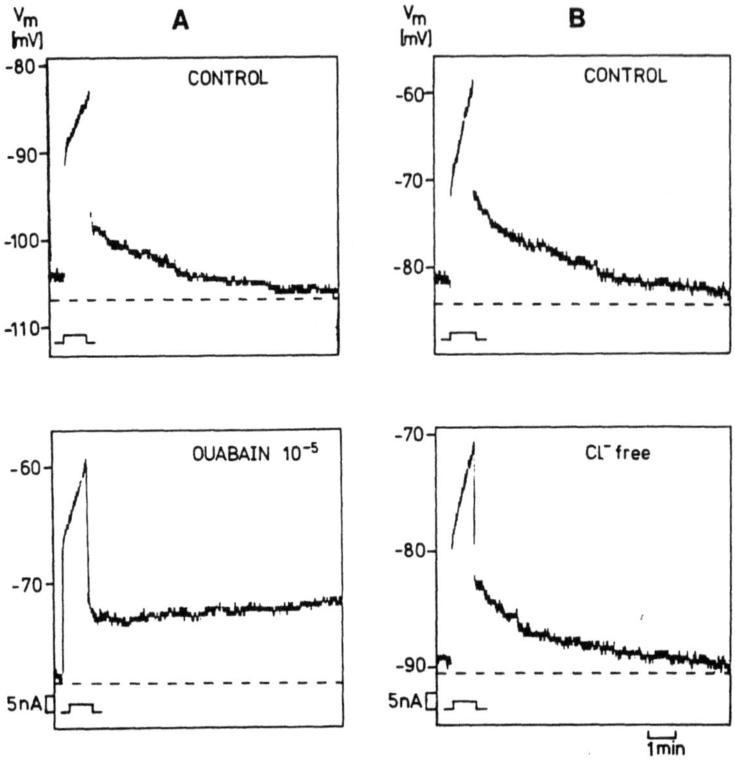

Fig. 12A,B. The effect of ouabain and chloride-free Ringer solution on recovery of the oligodendrocyte membrane potential after the intracellular injection of Na^+. The experimental set-up is described in Fig. 11. Recovery of the membrane potential (V_m) after the application of a 2.5 nA current pulse was influenced by ouabain (**A**), but remained unchanged after exposure to Cl^--free Ringer solution (**B**). (From Kettenmann et al. 1987)

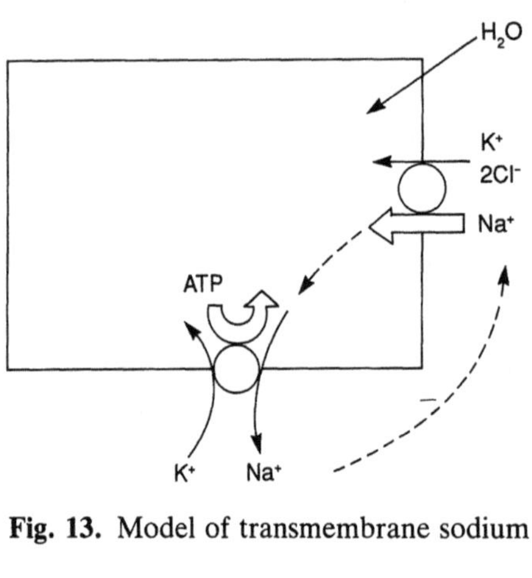

Fig. 13. Model of transmembrane sodium cycle. (From Walz and Hinks 1986)

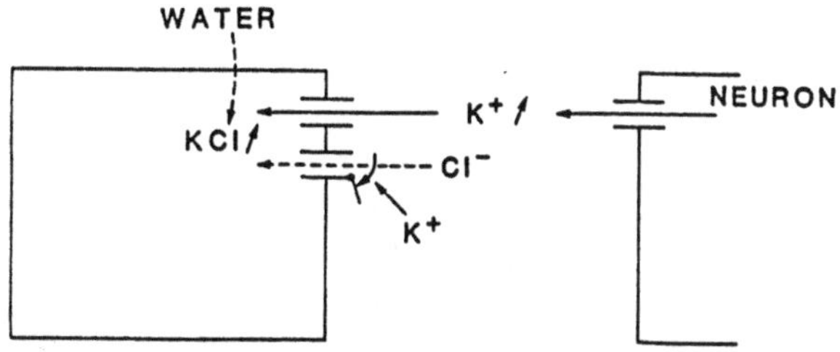

Fig. 14. Model of passive KCl uptake into glial cells, using a "modulated Boyle and Conway" mechanism. Voltage-gated Cl^- channels are opened indirectly by massive K^+ release from neurons via depolarization of membrane. The resulting KCl accumulation in glial cells would result in glial water uptake. (From Walz 1989)

3.3.2.3 Channel-Mediated KCl Uptake

Voltage-dependent Cl^- channels have recently been described in oligodendrocytes and astrocytes, leading to the suggestion that passive KCl uptake is involved in K^+ homeostasis (Bevan et al. 1985; Gray and Ritchie 1986; Sonnhof 1987; Nowak et al. 1987; Ballanyi et al. 1987; Barres et al. 1988). It has been suggested that the K^+ inward rectifier channel and the voltage-dependent Cl^- channel act together. Barres et al. (1988) found evidence of KCl uptake in cultured astrocytes and oligodendrocytes and termed it the "modulated Boyle and Conway mechanism". According to this mechanism, an increase in $[K^+]_e$ depolarizes the glia and opens Cl^- channels; KCl then enters the glial cells passively, accompanied by water (see Fig. 14), and is also driven out of the glia during the $[K^+]_e$ decrease. Both movements are effected by Donnan forces. In the course of this mechanism, K^+ is not transported from regions of high to regions of low concentration (i.e., by the spatial buffer mechanism) but is temporarily stored in glial cells which are in close proximity to the discharging neurones.

It was recently suggested by MacVicar (1984) and Quandt and MacVicar (1986) that an increase in $[K^+]_e$ which depolarizes the glia opens voltage-dependent Ca^{2+} channels, thereby creating an increase in intracellular free Ca^{2+}. The increase in $[Ca^{2+}]_i$ may open Ca^{2+}-dependent K^+ channels and in this way enhance the K^+-buffering capacity of the glial cells.

It can be concluded that glial cells play an important role in the maintenance of K^+ homeostasis in the CNS and receptor organs of invertebrates, insects, amphibians, and mammals, both during normal neuronal activity and in various pathological states of the CNS.

3.3.3 K^+ Diffusion in the ECS

Wherever K^+ accumulates in the ECS during neuronal activity, a K^+ gradient is formed, with the result that K^+ ions diffuse from regions with a high concentration to those with a lower concentration. The extracellular clefts are sufficiently wide (15–25 nm) to allow the unobstructed passage of hydrated K^+ (Kuffler and Nicholls 1966). K^+ ions, however, do not move in the ECS as they would in a free solution; the long-chain glycoproteins and glycolipids which are incorporated in the cell membranes and project into the intercellular clefts can carry negative charges which quite probably limit K^+ diffusion (Margolis and Margolis 1977; Nicholson 1980).

Since the movement of ions is influenced by the geometry of the ECS, it is obvious that K^+ diffusion in such a non-homogeneous and limited environment will be slowed down. Ion movements are affected by the geometry of the ECS in at least two ways including (a) the tortuosity (λ) of the intercellular clefts; and (b) the size of the ECS (the volume fraction, α) (Nicholson et al. 1979; Nicholson and Phillips 1981). Tortuosity increases the length of diffusion pathways. It is formally represented by the factor λ and the relationship between the diffusion coefficient in complex medium (D^*) and the diffusion coefficient in free solution (D) is $D^* = D/\lambda^2$ (Patlak and Fenstermacher 1975; Nicholson 1979). The factor α stands for the ratio of ECS volume (V_e) to total tissue volume (V)$\alpha = V_e/V$, i.e., the limited space available for the formation of extracellular ion gradients. These two factors, together with the previously described passive and active processes which participate in the reabsorption of K^+ into the cells and capillary bed, made it possible to apply Fick's law to the diffusion of K^+ in the ECS.

Studies using ISMs enable us to investigate ion diffusion in specific regions of the CNS in relation to time. The authors of the first diffusion studies with K^+-ISMs studied the diffusion of K^+ (Lux and Neher 1973; Fisher et al. 1976b; Pedley et al. 1976a), but this was soon found to be unsatisfactory, since K^+ ions are transported actively across cell membranes and thus also migrate through the intracellular medium. Other ions, not normally found in the CNS, are therefore used nowadays for diffusion studies and for studies to determine dynamic changes in the size of the ECS. The movements of monovalent cations have, for example, been studied by means of tetraethylammonium (TEA^+) or tetramethylammonium (TMA^+) ions, with the aid of K^+-ISMs (see Sect. 7.1; Nicholson et al. 1979; Nicholson and Phillips 1979, 1981). These studies show that if we incorporate factors α and λ in Fick's law, diffusion in the CNS can be described fairly satisfactorily. In an experimental situation the ions are introduced into the tissue by iontophoresis, i.e., the ions are released at a point with a constant flux for a finite period of time. The microelectrodes from which TEA^+ or TMA^+ is released iontophoretically act as the point source (Q) in $mol\,s^{-1}$:

$$Q = In/F$$

The appropriate solution of the diffusion equation is:

$$C(r, t) = \frac{Q\lambda^2}{4\pi D\alpha r} \operatorname{erfc}\left(\frac{r\lambda}{2\sqrt{Dt}}\right)$$

in which C is the concentration, I the iontophoretic current increment, n the transport number, F Faraday's constant, r the distance of the ISM from the ion source (i.e., from the iontophoretic pipette), D the diffusion coefficient, t the time, and erfc the complementary error function.

The diffusion parameters in the neuronal microenvironment can be determined using this equation. If we compare measurement in the brain with measurement in free solution, e.g., in dilute agar (where $\alpha = 1$ and $\lambda = 1$), we can determine the parameters in the diffusion equation (including α and λ) in a small tissue volume in vivo. Nicholson and Phillips (1981) found the values $\lambda = 1.48 \pm 0.05$ (mean $\pm$SE) and $\alpha = 21 \pm 5\%$ ($n = 59$) for TEA^+ movement in the rat cerebellum; the values for the movement of TMA^+ were similar – $\lambda = 1.49 \pm 0.05$ and $\alpha = 18 \pm 2\%$. When studied the movement of TMA^+ in the dorsal horns of the rat spinal cord we obtained similar values: $\lambda = 1.54 \pm 0.04$ and $\alpha = 24 \pm 1\%$ (Svoboda and Syková 1989, 1991). These and other measurements show that the ECS in the rat accounts for about one-fifth to one-quarter of the total brain and spinal cord volume. The diffusion coefficient for monovalent extracellular ions in nervous tissue is low, i.e., $\lambda^2 = 2.4$ (Nicholson and Phillips 1981). Similar λ values have been obtained in various parts of the brain using radiotracer methods (Levin et al. 1970; Patlak and Fenstermacher 1975), and radiotracer studies and impedance measurement have also yielded similar results on the size of the ECS (Van Harreveld 1972; Levin et al. 1970; Patlak and Fenstermacher 1975, Cragg 1979).

It can be concluded from these results that the movement of K^+ in the ECS cannot be characterized using the basic diffusion equation. In addition to the other processes previously mentioned, the redistribution of K^+ is also influenced by tortuosity and the size of the ECS. One of the important consequences of the change in size of the ECS which accompanies neuronal activity (see Chap. 7) could therefore be a change in the diffusion characteristic of K^+, and other ions, in nervous tissue.

3.3.4 K^+ Exchange Between Extracellular Fluid and Blood

It is unlikely that ion exchange between extracellular fluid and blood influences the course of dynamic ion changes which occur during neuronal activity. The time course for the exchange of K^+, which is largely an intracellular ion, would be extremely slow (half-time about 24 h). If K^+ is, however, present in the ECS in a very high concentration, and there is thus no exchange with intracellular K^+, the diffusion equilibrium between the extracellular fluid and the blood would depend on two main factors – the permeability of the capillary walls and the rate of blood flow in the vascular bed. Since capillary permeability for K^+ is about ten times greater than for Na^+ and Cl^- (Davson and Welch 1971), balanced diffusion would be attained in 15–30 min (Nicholson 1980). No such lengthy changes have been observed without the participation of intracellular reabsorption, however, and it is therefore generally acknowledged that exchange of K^+ between the blood and the extracellular fluid when $[K^+]_e$ is elevated is negligible. When $[K^+]_e$ rises to higher values, K^+ is immediately taken up by neurones and glial cells and is quickly redistributed. Measurement of the K^+ level in blood vessels of the cerebral cortex, sagittal sinuses, and cortical ECS of the cat during electrical stimulation and pharmacologically induced hyperexcitability showed that the transport of K^+ into the bloodstream had no physiological significance in the redistribution of accumulated K^+ ions in the ECS (Mutsuga et al. 1976).

4 Dynamic $[K^+]_e$ Changes

4.1 Dynamic $[K^+]_e$ Changes in the Spinal Cord

It was discovered in the first experiments using K^+-ISMs that $[K^+]$ in the ECS of the grey matter of the dorsal spinal horns and medulla rose by several mmol/l after stimulation of the sciatic nerve (Krnjević and Morris 1972; Vyklický et al. 1972). Since then, increases in $[K^+]_e$ have been described in a whole series of studies on the spinal cord of mammals and amphibians, in vivo and in vitro, after both electrical and adequate stimulation. The results of these studies, described in this chapter, can be summarized as follows:

- K^+ accumulates preferentially in the deep layers of the dorsal spinal horns, in the segments where afferent input is stimulated.
- K^+ is released from active neurones and primary afferent fibres.
- K^+ accumulation depends on the frequency, intensity, and duration of stimulation.
- K^+ which accumulates during stimulation is redistributed and cleared from the ECS chiefly through activation of the Na^+/K^+ pump in the neurones and by the glial cells.
- K^+ accumulated temporarily in the ECS depolarizes the neurones and nerve fibres in the vicinity, affects their excitability, and impairs the efficacy of spinal cord transmission.

4.1.1 $[K^+]_e$ Changes Induced by Electrical Stimulation of Peripheral Nerves

The maximum $[K^+]_e$ changes recorded using K^+-ISMs in the dorsal horns of the cat, the rat, and the frog after stimulation of peripheral nerves or spinal roots with a single electric pulse amount to 0.1–0.5 mmol/l (Figs. 17–19; Kříž et al. 1975; Czéh et al. 1981; Svoboda et al. 1988). Changes in $[K^+]_e$ are probably greater, but cannot be detected as they originate in the relatively large neurone population in the dorsal horns of the spinal cord which are too far away from the electrode. The electrode is therefore relatively far away from the nearest neurone (see Fig. 2) However, far greater changes were measured during repetitive stimulation. In mammals and amphibians, an increase from a resting $[K^+]_e$ of 3.0–3.5 mmol/l to 6–9 mmol/l was found (Figs. 4, 5A, 60, 61). In amphibians, greater changes than in mammals are usually observed, perhaps because active reabsorption and the activity of the Na^+/K^+ pump in these animals is less pronounced than in mammals. This is confirmed by the fact that the redistribution of accumulated K^+ in the frog is slower. In the frog, after termination of stimulation, we do not often see the drop in $[K^+]_e$ below the resting level, i.e., K^+ undershoot, which is typical in mammals. When K^+ undershoot is observed in the frog, it is more frequent in vivo than in vitro.

The increase in $[K^+]_e$ depends primarily on the frequency and intensity of stimulation. If low-threshold afferent fibres are stimulated supramaximally, however, the stimulation of high-threshold fibres no longer leads to increased changes in $[K^+]_e$ which are statistically significant. The frequencies at which the maximum increase in $[K^+]_e$ occurs range from 30 to 100 Hz. In mammals, the maximum increase – the ceiling level – is usually reached at lower electrical stimulation frequencies (30 Hz) than in amphibians (100 Hz). If the Na^+/K^+ pump is blocked by ouabain or naloxone (Syková et al. 1985a; Hájek et al. 1985), the K^+ ceiling rises in both mammals and amphibians, the redistribution of accumulated K^+ is slowed down, and K^+ undershoot below the resting level after termination of tetanic stimulation always disappears (Fig. 5; Krnjević and Morris 1975; Syková et al. 1985a; Syková 1976). K^+ is also redistributed more slowly during ischaemia, hypoxia, a drop in blood pressure, and hypothermia (Kříž et al. 1975; Syková et al. 1976). It is also the case that when the ceiling value for $[K^+]_e$ has been reached and stimulation continues, the ceiling value gradually falls (Figs. 4, 5) – probably because less and less K^+ accumulates and the Na^+/K^+ pump is continually activated, i.e., the steady state between the release and redistribution of K^+ is reached. It is also evident that the spatial buffer mechanism of the glial cells operates in this phase (Syková et al. 1988a; Walton and Chesler 1988; Jendelová and Syková 1991; Fig. 62).

K^+ undershoot presents convincing evidence of the redistribution of accumulated K^+ through activation of the Na^+/K^+ pump (see Sect. 3.3.1). Conversely, its presence and degree can furnish information on the amount of activity of the Na^+/K^+ pump. K^+ undershoot is, however, only observed in the middle of a population of activated neurones, i.e., at sites of maximum increase in $[K^+]_e$. Biochemical findings showing that naloxone inhibits Na^+/K^+ ATPase activity, while morphine or enkephalins stimulate such activity (Hájek et al. 1985), were confirmed by experiments using K^+-ISMs in which we demonstrated that naloxone inhibits the redistribution of K^+ and blocks the K^+ undershoot, whereas the administration of morphine or enkephalins is followed by an increase in K^+ undershoot (Syková et al. 1985a).

4.1.2 Depth Profile of $[K^+]_e$ Changes in the Spinal Cord

The increase in $[K^+]_e$ during stimulation of afferent input in the spinal cord of mammals and amphibians has a characteristic time course. An increase in $[K^+]_e$ is observed in practically all layers of the dorsal spinal horns, but the maximum changes are always found in the deeper layers. In mammals (rats, cats, mice, and golden hamsters), we observed the maximum change in Rexed's laminae III–VII (see Fig. 15) and in Ebbesson's lateral fields in the frog (Ebbesson 1976; Figs. 16 and 17).

Activated neurones in the deeper layers of the dorsal horns are the chief source of K^+ which accumulates in the ECS during stimulation. However, 10%–35%

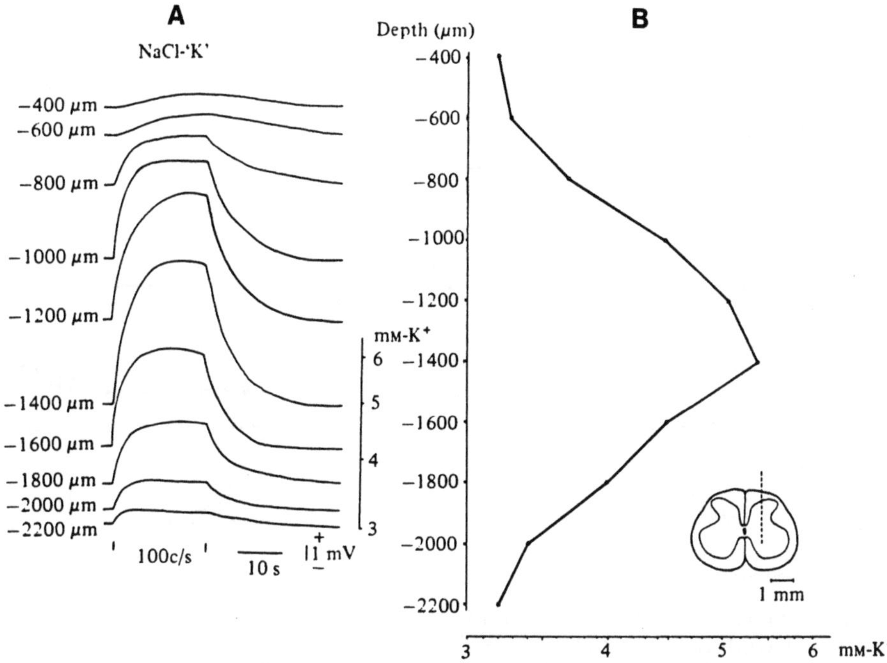

Fig. 15. **A** Elevation of $[K^+]_e$ at different depths in segment L7 of the cat spinal cord during tetanic stimulation of the posterior tibial nerve (100 Hz, 15 s). **B** Distribution of $[K^+]_e$ changes at different depths during penetration of the cord by the microelectrode (*bottom right*). (From Kříž et al. 1974)

of the K^+ released in the frog's dorsal spinal horns comes from the primary afferent fibres themselves. In experiments where synaptic activity was blocked by the addition of 20–30 mmol/l $MgSO_4$ or 2–4 mmol/l $MnCl_2$ to the Ringer solution used for perfusion of the isolated frog spinal cord, the increase in $[K^+]_e$ after electrical and adequate stimulation amounted to only 10%–35% of the increase obtained if synaptic activity was not blocked (see Figs. 5 and 62; Syková and Vyklický 1977a; Nicoll 1979; Davidoff et al. 1988).

At the site of activated interneurones, in the deeper layers of the dorsal spinal horns of both mammals and amphibians, the post-stimulation increase in $[K^+]_e$ is three to ten times greater than in the superficial layers. The $[K^+]_e$ increment in the ventral spinal horns is also small since, unlike interneurones, stimulated motoneurones cannot follow a high stimulation frequency. The number of cells per mm^3 in the ventral horns is also smaller than in the deeper layers of the dorsal horns (Aitken and Bridger 1961). Even antidromic stimulation of the motoneurones during electrical stimulation of the ventral spinal roots did not produce changes in $[K^+]_e$ in the ventral horns of more than 0.1 mmol/l (Fig. 16). During stimulation of descending pathways, however, when interneurones in the ventral spinal horns are activated, a different distribution of $[K^+]_e$ changes is observed (Czéh et al. 1981).

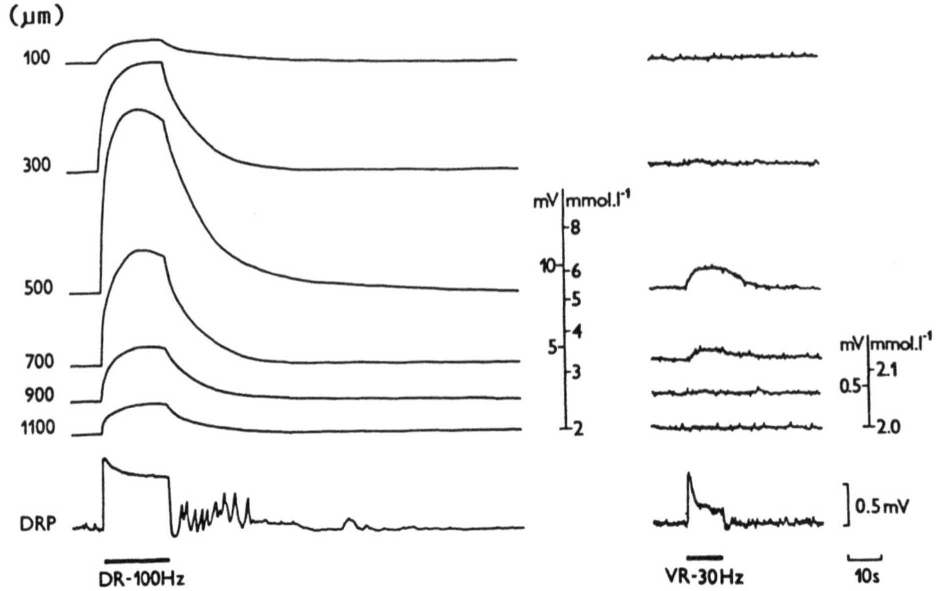

Fig. 16. Comparison of the increase in $[K^+]_e$ at different depths of segment IX of the frog spinal cord during dorsal root (*left*) and ventral root (*right*) stimulation. The K^+-ISM was introduced from the dorsal surface of the cord. Dorsal root potentials were recorded during each stimulation, but only those accompanying recordings from a depth of 500 μm are illustrated. (From Czéh et al. 1981)

4.1.3 Electrical Stimulation of Descending Pathways

In experiments where descending spinal pathways of the frog were electrically stimulated, $[K^+]_e$ increased substantially in the ventral horns (Czéh et al. 1981; Fig. 17). Irrespective of whether we studied the isolated spinal cord, or the spinal cord in situ, we stimulated the ventrolateral columns at medulla oblongata level. Since the increase in $[K^+]_e$ is proportional to the number and density of activated neurones and to the frequency of their action potentials (Syková et al. 1974; Kříž et al. 1974; Czéh et al. 1981), the increase in $[K^+]_e$ was obviously the result of repeated activity of interneurones in the ventral spinal horns. The changes observed in $[K^+]_e$ were similar to those seen in the dorsal spinal horns during peripheral nerve stimulation. After stimulation of the ventrolateral columns with a single electric impulse, the maximum increase in $[K^+]_e$ was about 0.15 mmol/l, whereas it rose to 9–10 mmol/l during tetanic stimulation. In simultaneous tetanic stimulation of descending pathways and antidromic stimulation of ascending fibres in the dorsal spinal columns (i.e., when the dorsal quadrant of the cervical cord was not removed prior to electrical stimulation), the increase in $[K^+]_e$ amounted to 12.5–15.0 mmol/l and thus exceeded the ceiling level for the increase in $[K^+]_e$ in the dorsal spinal horns (Czéh et al. 1981).

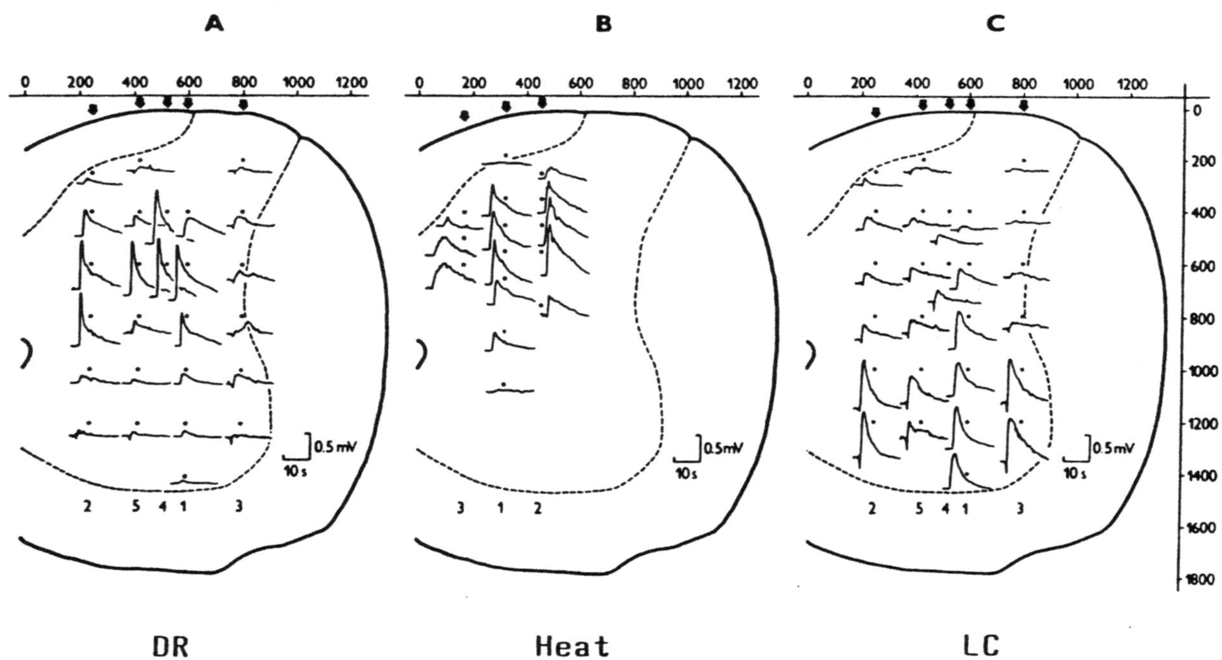

Fig. 17A–C. Distribution of $[K^+]_e$ changes in segment IX of the frog spinal cord. **A** The $[K^+]_e$ increment induced by application of a single electric pulse to the skin of a hind limb. The recordings are all taken from the same experiment, using the same microelectrode, which was introduced five times, from the dorsal surface, into the spinal cord in situ (the penetrations are numbered 1–5 in the order in which they were performed). The depth at which $[K^+]_e$ changes were recorded is marked with a *dot*. *DR*, dorsal root. **B** The increase induced in $[K^+]_e$ by applying hot water for 1–2 s to the skin of a hind limb. **C** The increase in $[K^+]_e$ induced by stimulating the ventrolateral columns (*LC*) with a single electric pulse. (From Czéh et al. 1981)

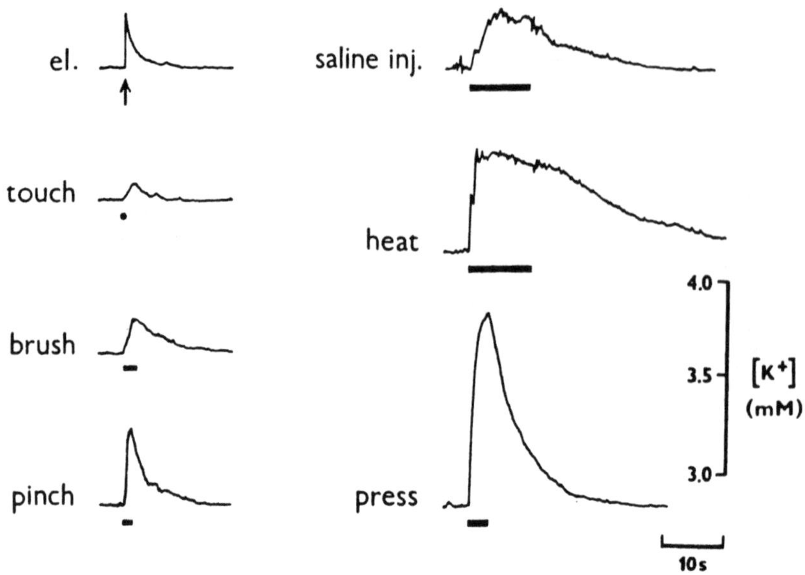

Fig. 18. Transient increases in $[K^+]_e$ in the L_4 spinal cord segment of the rat in response to adequate stimuli applied to the plantar region of the ipsilateral hind paw. The K^+-ISM was inserted to a depth of 600 μm from the dorsal surface. The curves represent responses to a single electrical stimulus applied through acupuncture needles inserted below the plantar surface (*el.*), to light touch, brush, brief pinch, saline injection, application of about 1 ml of hot water, and strong pressure to the toes. *Horizontal bars* indicate stimulus duration. (From Svoboda et al. 1988)

4.1.4 $[K^+]_e$ Changes Induced by Adequate Stimulation

4.1.4.1 Acute Nociceptive and Non-Nociceptive Stimuli

$[K^+]_e$ changes were found in the dorsal spinal horns after adequate, as well as electrical stimulation (Syková et al. 1980; Czéh et al. 1981; Svoboda et al. 1988; Heinemann et al. 1990). In the frog (Figs. 17 and 19), the rat (Fig. 18), and the cat, $[K^+]_e$ in the dorsal spinal horns rises by several tenths of a mmol/l after tactile stimulation of the appropriate receptor field on the skin of the hind limb. If the tactile stimuli are repeated in quick succession, the summation of $[K^+]_e$ changes is similar to that seen after electrical stimulation. Acute nociceptive stimuli which do not lead to irreversible tissue damage produce greater $[K^+]_e$ changes than tactile stimuli or single, transcutaneously applied electric pulses (Figs. 18 and 19). A painful pinch of one of the joints of the hind toes with forceps, pressing a paw down on a board with a rod, or heat stimuli (radiant heat, a drop of hot water), of several seconds' duration, lead to a $[K^+]_e$ increment of 0.1–1.0 mmol/l. The prolongation or quick repetition of nociceptive stimulation leads to summation of the changes observed in $[K^+]_e$ until the maximum change – the point at which the relevant skin receptors have

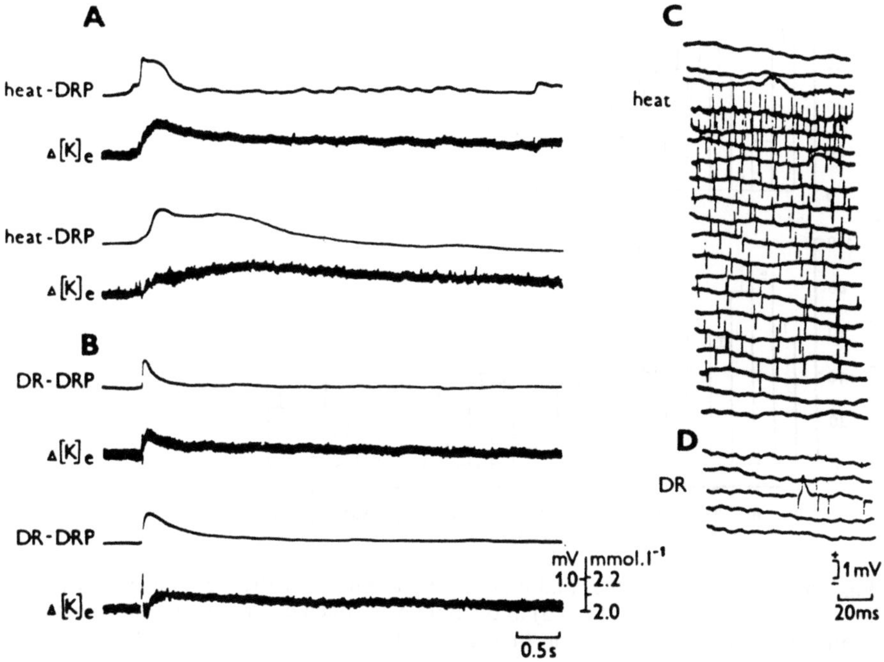

Fig. 19A–D. The time course of changes in $[K^+]_e$ and of dorsal root potentials (*DR-DRP*) in segment IX of the frog spinal cord. **A** Responses to the application of hot water to the skin of a hind limb (1–2 s, 58°C). **B** Responses to the stimulation of the skin with a single electrical pulse. The K^+-ISM was introduced to a depth of 600 μm. (The recordings were always made at two speeds.) **C**, **D** Characteristic extracellular recordings of action potentials evoked in an interneurone at a depth of 600 μm in the same frog by the application of hot water (**C**) and a single electric pulse (**D**). (From Czéh et al. 1981)

adaptated to the given stimulus – is attained. The recording of extracellular action potentials of neurones in the dorsal spinal horns during these experiments showed that $[K^+]_e$ changes correlate well with the degree and duration of evoked activity of the neuronal population concerned (Czéh et al. 1980, 1981). Figure 19 shows a typical record of dorsal root potentials (DRPs), changes in $[K^+]_e$ and extracellular activity from a single frog dorsal horn neurone in situ after transcutaneous stimulation with a single electric pulse, and the application of 1 ml hot water (60°C) to the skin of the ipsilateral hind limb. Increased activity was greatest and longest after the application of hot water to the appropriate skin receptor field; the increase in $[K^+]_e$ was also greatest. An increase in $[K^+]_e$ of approximately 50% was also obtained by electrical or adequate stimulation of the same site on the contralateral hind limb (Czéh et al. 1980).

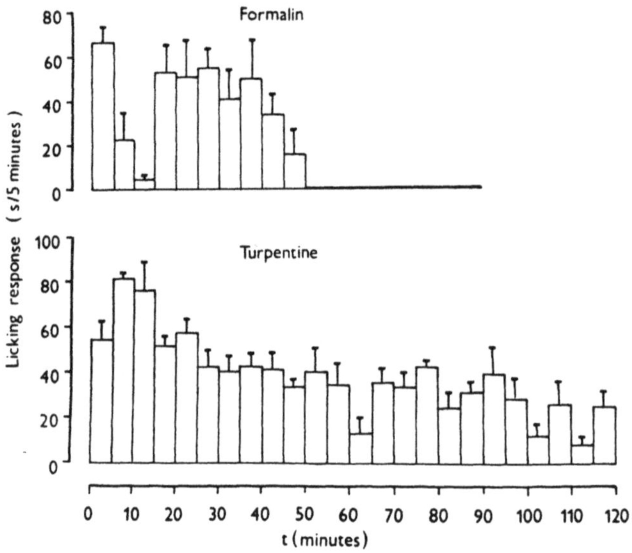

Fig. 20. Time course of the hind limb licking in rats after the s.c. injection of formalin ($n = 6$) and turpentine oil ($n = 5$) into a hind paw. Each column represents the duration of the licking response per 5 min (mean ± SE). (From Svoboda et al. 1988)

4.1.4.2 Chronic Nociceptive Stimuli

In the rat, chronic nociceptive stimulation produces a long-lasting increase in $[K^+]_e$ and disturbance of local ionic homeostasis in the dorsal spinal horns (Svoboda et al. 1988). In our studies chronic nociceptive stimulation was represented by the application of algogens (pain-producing chemicals) or by local burns which caused chronic skin lesions with an extensive local inflammatory reaction. This type of nociceptive stimulation differs from phasic nociceptive stimulation by electric shock, pinching, radiant heat, or a drop of hot water, and imitates well the chronic pain associated with various kinds of illnesses (see Handwerker 1980). The behaviour of a group of rats treated with algogens was first studied. Figure 20 shows the time course of the licking response, to the painful sensation caused by the formalin test (the s.c. administration of 0.1–0.5 ml 5% formalin), or the s.c. injection of turpentine (0.1–0.5 ml turpentine oil) into the planta of a hind limb. Study of the animals' behaviour after the injection of algogens clearly documents the nociceptive reaction. Rats generally hold the limb flexed, do not put it on the ground, and often lick it intensively. This licking is a relatively accurate indicator of the intensity and duration of the painful reaction (Hunskaar et al. 1985; Sugimoto et al. 1986; Svoboda et al. 1988). Electrophysiological studies have shown that both the injection of the substances, and the peripheral lesions producing acute inflammation, cause an instantaneous and permanent increase in the excitability of neurones in the marginal and deeper layers of the dorsal spinal

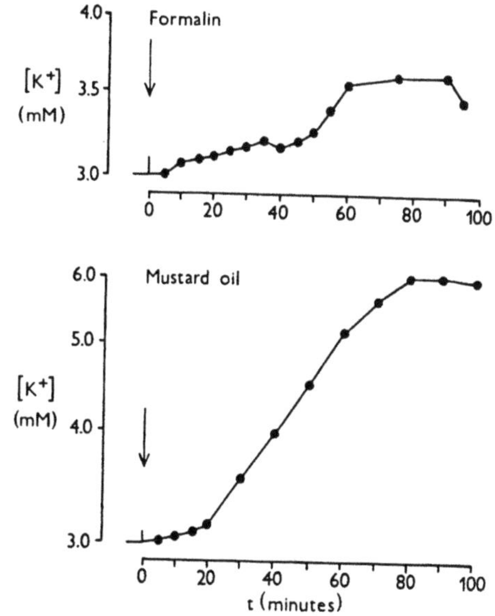

Fig. 21. The increase in $[K^+]_e$ induced in rat dorsal horns by the s.c. injection of 0.3 ml formalin into a hind paw or by the application of mustard oil to the skin of a paw. Each *point* represents the increase in $[K^+]_e$ from baseline, i.e., the level of $[K^+]_e$ before injury. (From Svoboda et al. 1988)

horns, with a frequently continuous, or sometimes burst-like, incidence of action potentials (Besson and Chaouch 1987; Banna et al. 1986).

In rats, the administration of algogens (formalin, turpentine) or the irritant allyl isothiocyanate (mustard oil) produced an increase in $[K^+]_e$ which persisted for at least 50 min after treatment, and often longer (Svoboda et al. 1988). The degree and time course of the changes varied, the increase ranging from 0.3 to 3.0 mmol/l, i.e., from a resting 3.0 mmol/l to 3.3–6.0 mmol/l (Figs. 21 and 22). The increase in $[K^+]_e$ was accompanied by hyperexcitability of the spinal neurones, so that subsequent phasic nociceptive stimuli (pinching, pressure) and tactile stimuli evoked $[K^+]_e$ changes and associated field potentials of up to double the original size (Fig. 22). A similar increase in $[K^+]_e$ and excitability was also observed in the case of peripheral lesions produced by burns. The application of 2–3 ml boiling water (80°–90°C) or hot liquid paraffin (120°–150°C) first produced an acute reaction, i.e., a sudden transient increase in $[K^+]_e$, followed by a return to the resting level. It was not unitl 5–20 min after the application of heat stimuli that a reaction to chronic injury, similar to that produced by the injection of algogens, began slowly to make its appearance. In some cases, the initially slow elevation of $[K^+]_e$ was followed by a sudden further increase of more than 2–3 mmol/l, while in others this relatively large increase in $[K^+]_e$ did not occur until the damaged area had been further stimulated by a tactile stimulus or by pinching. If the $[K^+]_e$ reached about 6 mmol/l or over, however, we often observed a decrease in excitability, i.e.,

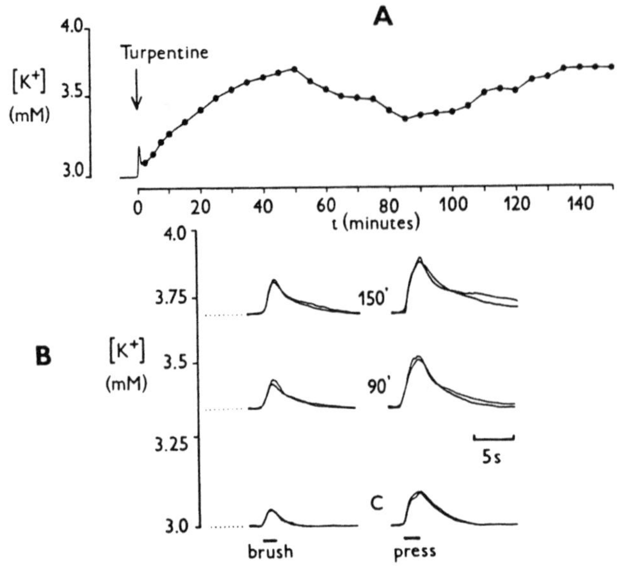

Fig. 22A,B. The increase in $[K^+]_e$ in rat dorsal horns at a depth of 500 μm after the s.c. injection of 0.3 ml turpentine oil into a hind paw. **A** Each *point* corresponds to an increase in $[K^+]_e$ from the resting level before injury. **B** The transient increase in $[K^+]_e$ evoked by brush or light pressure to the toes. *C*, control responses before injecting turpentine; *90′* and *150′*, enhanced responses 90 and 150 min after injection. (From Svoboda et al. 1988)

both adequate and electrical stimuli induced shorter and smaller changes in $[K^+]_e$ and associated field potentials decreased. In some cases the increase in K^+ was of an epileptiform character, i.e., $[K^+]_e$ rose suddenly to over 10 mmol/l and than returned slowly to values in the region of 6 mmol/l.

The resting $[K^+]_e$ level at different depths in the spinal cord varies according to the level of spontaneous activity (see Sect. 4.1.5). In the rat spinal cord, we found that $[K^+]_e$ in the deeper layers of the dorsal horns was about 0.5 mmol/l higher (i.e., it reached values of 3.5 mmol/l) than in the superficial layers or in the ventral spinal horns. The administration of algogens or the application of heat stimuli raised resting $[K^+]_e$ in substantially larger areas of the spinal dorsal horns (Fig. 23).

These experiments show that chronic nociceptive stimuli which lead to hyperexcitability and abnormal activity of the interneurones in the dorsal spinal horns, are accompanied by significant ionic changes in the ECS (also see Sect. 6.2). The increase in $[K^+]_e$ was, however, blocked when the application of the nociceptive stimulus was preceded by local procaine anaesthesia. Nevertheless, in experiments where abnormal activity had already developed, and $[K^+]_e$ had already risen, a procaine block was ineffective. These results provide evidence that the prolonged elevation of the activity of C fibres after the administration of algogens or after injury, evoked in the dorsal spinal horns a process (abnormal activity), which, once it had started, became self sustaining and independent of peripheral input (Woolf 1983; Svoboda et al. 1988). The permanent elevation of the resting $[K^+]_e$ in the dorsal horns which accompanies this activity in chronic painful states must have a significant effect on spinal transmission and on the transmission of nociceptive information to the brain. It is most probably not just a concomitant phenomenon of negligible importance; consideration should be given to the possibility that the accumulation of K^+ in the microenvironment of neurones and glial cells has a functional significance, evidenced by the modulatory effect on neuronal excitability, the depolarization of primary afferents, and synaptic transmission, etc., as discussed in Sect. 4.3.

4.1.5 $[K^+]_e$ Changes Associated with Spontaneous Activity in the Dorsal Horns of the Spinal Cord

The resting $[K^+]_e$ in the brain and spinal cord varies at about 3.0 mmol/l. Using K^+-ISMs, values of 2.6–3.8 mmol/l were measured in various structures of the brain and spinal cord. Different values probably result from different levels of spontaneous neuronal activity; the type of anaesthesia, drugs, hypoxia, or local tissue damage in the course of the experiment also contributes to these differences (Krnjević and Morris 1972; Vyklický et al. 1972; Vyskočil et al. 1972; Lux and Neher 1973; Prince et al. 1973; Futamachi et al. 1974; Svoboda et al. 1988). Interspecies differences in the resting $[K^+]_e$ also exist. As far as correlation of the resting level to the spontaneous activity level in some of the structures of the CNS is concerned, any differences in $[K^+]_e$ which develop in this way may play a physiologically important role, since they could influence the excitability of neuronal elements and the transmission of impulses in a given structure. Variations in the mesencephalic reticular formation (MRF) during spontaneous activity amount to a few tenths of a mmol/l (see Sect. 4.2.2). Satoh et al. (1979) showed that similar variations also occurred in the MRF during rapid eye movement (REM) sleep.

The interneurones in the dorsal spinal horns of mammals and amphibians often have a high spontaneous activity level and $[K^+]_e$ therefore varies. We found that $[K^+]_e$ in the submeningeal fluid was close to 3.0 mmol/l and that the resting level in the superficial layers of the dorsal spinal horns and in the ventral spinal horns was hardly any different. In the deeper layers of the dorsal horns and in the intermediate region, however, the resting $[K^+]_e$ was 0.5–1.5 mmol/l higher (Figs. 23 and 24) and varied according to the level of spontaneous activity, the temperature, hypoxia, and anaesthesia (Syková et al. 1983; Svoboda et al. 1988). In experiments on the isolated frog spinal cord it was demonstrated that the differences in the resting level according to depth were actually the outcome of spontaneous activity of interneurones in the dorsal horns, since they disappeared after a block of synaptic transmission or temperature changes (procedures which prevented spontaneous activity) (Fig. 24; Syková et al. 1983).

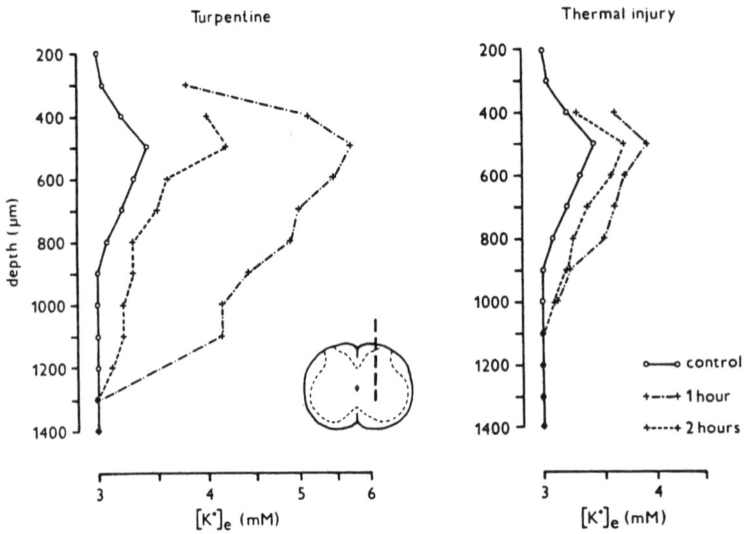

Fig. 23. The effect of the s.c. injection of turpentine and of thermal injury on $[K^+]_e$ at various depths in segment L_4 of the rat spinal cord. Control – the $[K^+]_e$ depth profile before injury. It should be noted that $[K^+]_e$ is higher in the deeper layers (400–700 μm) of the dorsal horns. The figure depicts the elevation of $[K^+]_e$ by 3.0 mM and 1.0 mM, 1 h after application of turpentine and thermal injury, respectively. There was partial recovery to the preinjury levels 2 h after injuries. (From Svoboda et al. 1988)

4.1.6 $[K^+]_e$ Changes Induced by Systemic Administration of Drugs, Transmitters, and Neuropeptides

The administration of certain amino acids causes an increase in $[K^+]_e$. The application of GABA (10^{-3} mol/l) to the dorsal root ganglia, sympathetic ganglia, and vagus nerve of the rat is associated with a $[K^+]_e$ increment of 0.5–1.0 mmol/l (Deschenes and Feltz 1976; Förstl et al. 1982). Similarly, the application of GABA to the isolated frog spinal cord in a concentration of 10^{-4} mol/l or more caused K^+ to be released from the cells and produced a 0.5–1.0 mmol/l increase in $[K^+]_e$ (Kudo and Fukuda 1976; Syková 1978). This increase is probably the outcome of secondary K^+ release, since the primary effect of GABA is known to increase membrane permeability for Cl^- (Krnjević 1974). A shift of one of the cations must evidently occur during the release of Cl^- to ensure that electroneutrality is preserved. The most likely cation is K^+, firstly because neuronal membranes in the resting state are readily permeable to K^+ and secondly because K^+ is the main intracellular cation. An increase in $[K^+]_e$ was also found if GABA was administered after blocking synaptic activity by adding 20 mmol/l $MgSO_4$ to the Ringer solution, or substituting Mg^{2+} for Ca^{2+} (Kudo and Fukuda 1976; Syková and Vyklický 1977b).

An increase in $[K^+]_e$ in the isolated frog spinal cord was also observed after the administration of glutamate (an excitatory transmitter), although the mech-

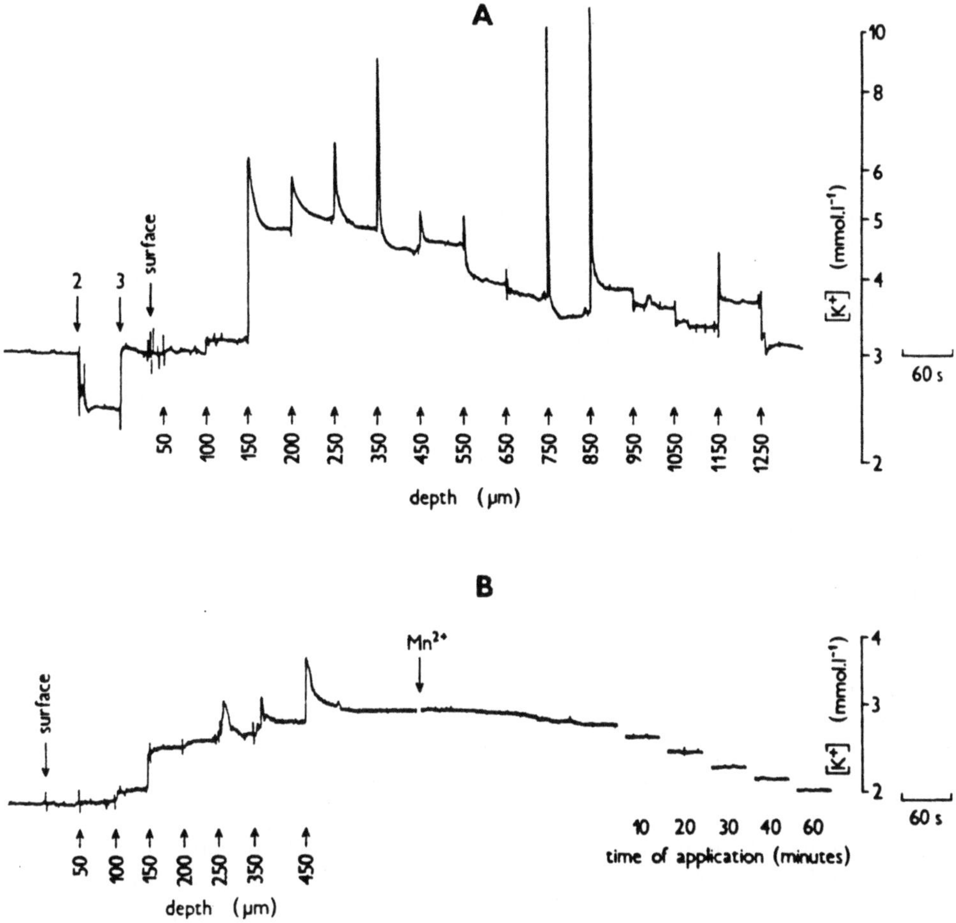

Fig. 24A,B. The [K$^+$]$_e$ at various depths of the unstimulated frog spinal cord. **A** Experiment on spinal cord of spinalized frog (spinal cord transsected at cercical level). The *arrows* marked *2* and *3* indicate the application of Ringer solution either with 2 or 3 mmol/l [K$^+$]$_e$ to the surface of the spinal cord; the K$^+$-ISM placed above this surface was, therefore, immersed in Ringer solution. The *other arrows* indicate micromanipulator movements and the *numerals* the microelectrode depth below the dorsal surface of the spinal cord. **B** The same experiment, but on the isolated frog spinal cord superfused with Ringer solution containing 2 mmol/l K$^+$. At a depth of 500 μm the Ringer solution with 2 mmol/l $MnCl_2$ was applied and the decrease in [K$^+$]$_e$ is shown at 10-min intervals. The peaks in **A** and **B** were caused by microelectrode penetration of the cells and had rapid decay time. The value of [K$^+$]$_e$ following these peaks remains constant and represents the actual resting [K$^+$]$_e$. (From Syková et al. 1983)

anism of action is different to that of GABA (an inhibitory transmitter) (Sonnhof and Buhrle 1980; Vyklický et al. 1982). The K$^+$ efflux evoked by the administration of certain excitatory amino acids could, however, be blocked by specific substances, or by using mono- or diaminodicarboxylic acid to block receptor activation (Hackman et al. 1982).

The local application of serotonin (5-hydroxytryptamine) produced an increase in $[K^+]_e$ in hippocampal slices (Segal and Gutnick 1980), accompanied by hyperpolarization of the neurones and cessation of their spontaneous spike discharge. Methysergide, an antagonist of serotonin, inhibited these effects. The authors suggested that serotonin hyperpolarizes hippocampal cells by activating Na^+- and Ca^{2+}-independent K^+ channels, with a resultant increase in $[K^+]_e$.

Lithium salts are commonly used in the treatment of psychiatric disorders such as mania and recurrent depression. Superfusion of the cerebral cortex with a solution having the same composition as the CSF and containing Li^+, or the administration of a diet containing LiCl, raised the $[K^+]_e$ level in rats (Grafe et al. 1983). The post-stimulation K^+ undershoot below the resting level was smaller after superfusion with the solution containing Li^+. It has been suggested that Li^+-related changes in $[K^+]_e$ result from the inhibition of the Na^+/K^+ pump.

Perfusion of the isolated frog spinal cord with Ringer solution containing naloxone raises the resting $[K^+]_e$ level and depolarizes the dorsal spinal roots, as does the administration of ouabain (Syková et al. 1985a; Fig. 5). The administration of leucine enkephalin (LE), methionine enkephalin, or morphine reduces the resting $[K^+]_e$ level and causes hyperpolarization of primary afferents (Syková et al. 1985b; Suzuki et al. 1986). The application of naloxone and ouabain caused the transient changes induced in $[K^+]_e$ by stimulation to increase, as did the DRPs. When stimulation was discontinued, the accumulated K^+ was redistributed more slowly and the K^+ undershoot disappeared. The administration of an enkephalin or morphine, however, had the opposite effect. Experiments on the membrane fraction of the frog spinal cord showed that both morphine and enkephalins, in 10^{-7}–10^{-4} mol/l concentration, stimulated Na^+/K^+ ATPase activity (Hájek et al. 1985). Administration of the specific antagonist naloxone had already inhibited Na^+/K^+ activity in low concentrations (10^{-7} mol/l). It can be thus concluded from these results that the $[K^+]_e$ increment observed after administration of naloxone is due to inhibition of the Na^+/K^+ pump.

The presence of various other transmitters and neuropeptides, their agonists, antagonists, and other drugs in the dorsal horns can also have specific and nonspecific effects on the $[K^+]_e$ level. Likewise, an increase in $[K^+]_e$, with all its consequences for the excitability of neuronal elements, transmitter release, synaptic activity, and spontaneous activity (see Sect. 4.3), can result in serious side effects as it obscures or accompanies the specific action of the aforementioned substances.

4.2 Dynamic $[K^+]_e$ Changes in the Brain

The introduction of ISMs into neurophysiological research has also made it possible to measure directly dynamic changes in the ionic composition of the extracellular and intracellular space in the brain. Experiments have shown that

$[K^+]_e$ in the brain changes during physiological neuronal activity as well as in various pathological states, many indicating that transient $[K^+]_e$ changes are associated with spontaneous activity of the neurones in specific brain structures and that significant transient changes occur during sensory stimulation. Protracted $[K^+]_e$ changes, in extreme cases irreversible, have been found during epilepsy, spreading depression, anoxia, and treatment with drugs (for review see Syková 1983; Sect. 4.3).
Over the past ten years dynamic changes in $[K^+]_e$ in the cerebral cortex, cerebellum, MRF, lateral geniculate nucleus, and hippocampus have been studied. As is the case in the spinal cord, the resting $[K^+]_e$ in the brain is either the same as, or higher than, that in the CSF. K^+ activity measured with ISMs in the non-stimulated brain (i.e., during neuronal activity which can be assumed to be spontaneous) varies from 2.6 to 5.0 mmol/l. Again, as in the spinal cord, these relatively marked differences in the resting $[K^+]_e$ are associated not only with different spontaneous activity levels in specific parts of the brain (Fig. 25), but also with interindividual and interspecies differences, and with changes in the experimental conditions. Dynamic $[K^+]_e$ changes in the brain also follow a very similar time course to the changes in the spinal cord which have been described in Sect. 4.1.

4.2.1 Dynamic $[K^+]_e$ Changes in the Cerebral Cortex and Striatum

During electrical stimulation of afferent input (e.g., of the nucleus ventralis posterolateralis thalami), $[K^+]_e$ in the mammalian cerebral cortex rises by up to as much as 7 mmol/l (Heinemann and Lux 1977). It does not, however, rise any further – not even during direct stimulation of the cerebral cortex (Futamachi et al. 1974; Lothman et al. 1975) – because it has reached the ceiling level described in Sect. 4.1. Greater changes occur only if epileptiform activity or spreading depression develops, or in the presence of severe ischaemia or anoxia (for reviews see Prince 1978; Bureš and Burešová 1981; Syková 1983). The ceiling level can vary according to experimental conditions. It is higher in young animals, e.g., in the striatum of 1- and 2-day-old chicks (Fig. 26A; Jendelová et al. 1988; Syková et al. 1990), and in immature young rats (Urban et al. 1985). As in the spinal cord, the ceiling value in the cerebral cortex also falls during stimulation lasting longer than 10–20 s. The post-stimulation $[K^+]_e$ undershoot below the normal resting level points to active K^+ reabsorption by means of the K^+-stimulated Na^+/K^+ pump (Fig. 26). If the reabsorption of K^+ is impaired – e.g., due to anoxia or other metabolic disturbances – or if the ion permeability of the neuronal and glial membranes is raised, as is the case in spreading depression, the ceiling level rises owing to impaired K^+ reabsorption (Vyskočil et al. 1972; Prince et al. 1973; Mutani et al. 1974; Lothman et al. 1975; Sugaya et al. 1975).
As seen in the spinal cord, neuronal activity evoked in the brain by adequate stimulation is accompanied by changes in $[K^+]_e$. In the visual cortex of the cat (Fig. 27) and in the ectostriatum of 2-day-old chicks (Fig. 28), a rise in $[K^+]_e$

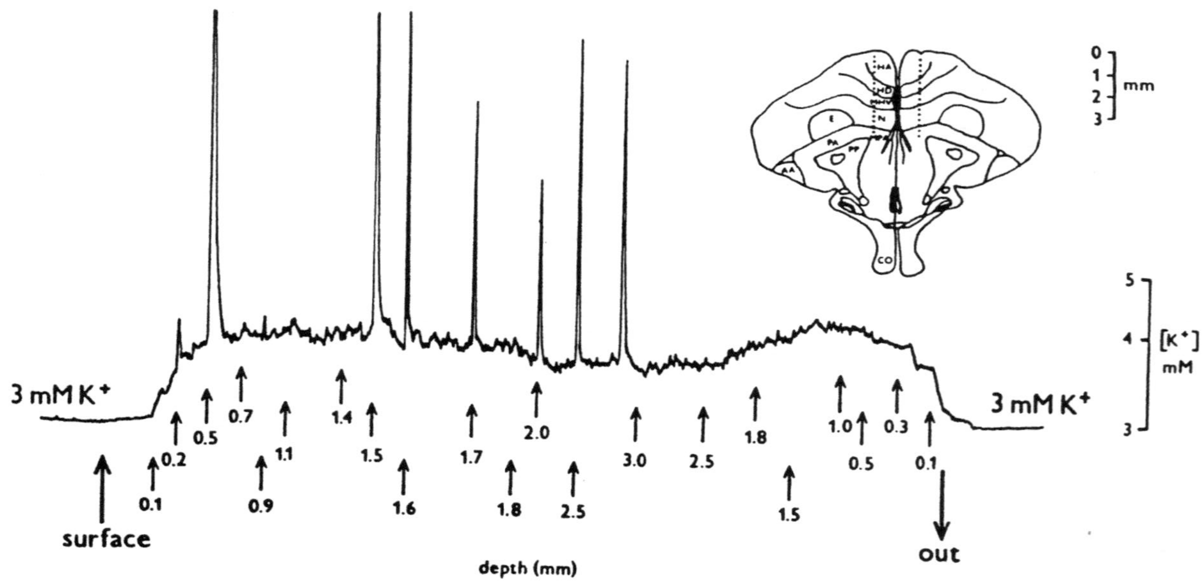

Fig. 25. The resting $[K^+]_e$ in the forebrain of a 2-day-old chick. Penetration of the brain by K^+-ISM is illustrated schematically (*top right*). Saline with 3 mmol/l K^+ was applied to the surface of the brain and $[K^+]_e$ was measured at different depths on the way into and out of the brain. The depth is denoted by *arrows* and a *number* representing the depth in mm. At all depths, from 0.5 to 3.0 mm, the resting $[K^+]_e$ varied from 3.5 to 4.0 mmol/l K^+. Large peaks, which were recorded when the microelectrode was moved in the brain, are due to the K^+ release from damaged cells. (After Syková et al. 1990)

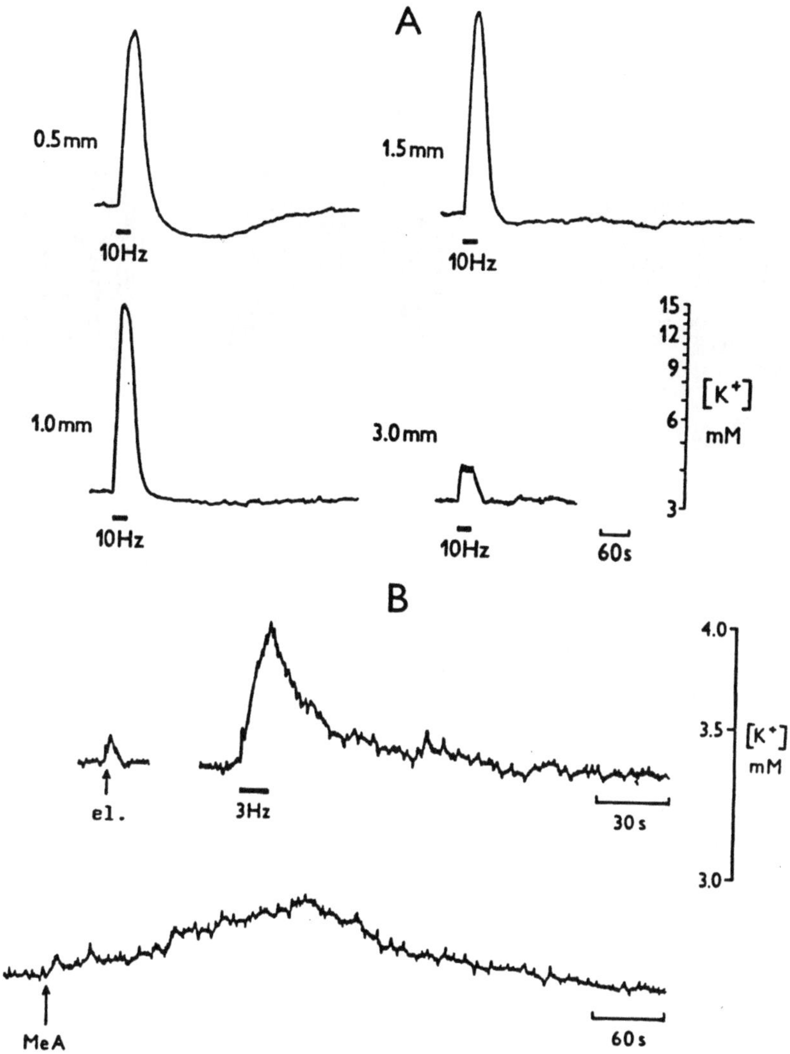

Fig. 26A,B. Stimulation-induced increase in $[K^+]_e$ in the brain of a 2-day-old chick. **A** Tetanic electrical stimulation of the surface of the brain with a bipolar platinum electrode (10 Hz, inter-electrode distance 5 mm). These recordings were made at depths of 0.5, 1.0, 1.5, and 3.0 mm below the surface of the brain (see Fig. 21). **B** Recordings of the increase in $[K^+]_e$ induced at a depth of 3.0 mm after stimulating the beak with a single electric pulse (*el.*), after repeated electrical stimulation at 3 Hz, and after administering the taste aversive substance methyl anthranilate (*MeA*). (From Syková et al. 1990)

accompanies evoked neuronal activity during stimulation of the receptive field with visual stimuli (Singer and Lux 1975; Syková et al. 1990). A $[K^+]_e$ increase of 0.2–0.5 mmol/l was observed in the striatum of 1- and 2-day-old chicks after repetitive stimulation of their beak with electric pulses at a frequency of 3 Hz,

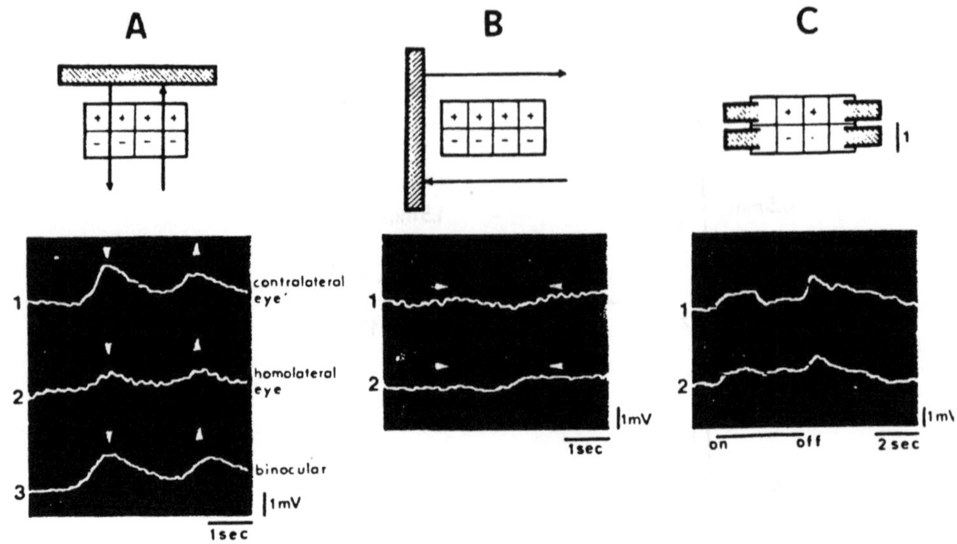

Fig. 27A–C. K^+ signals in response to moving light stimuli in the cat striate cortex. The graphs above the traces indicate the respective stimulation conditions. The receptive field covers a cell recorded 50 μm above the K^+ signals. All K^+ signals are averaged from 10 stimulus presentations. *1, 2, 3*, changes in $[K^+]_e$ in response to a horizontally oriented light slit which moves up and down over the field (*arrow on traces*). The stimulus is presented either to the contralateral eye (**A**), the homolateral eye (**B**), or both eyes simultaneously (**C**). (From Singer and Lux 1975)

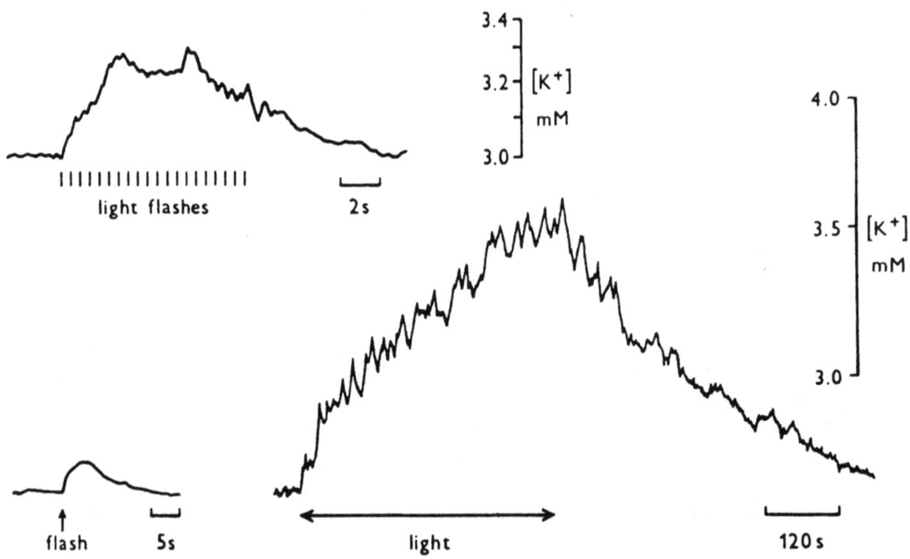

Fig. 28. Increase in $[K^+]_e$ evoked in the ectostriatum of 2–3-day-old chick (5 mm proximally from the ear bar axis, 3.5 mm laterally from the midline) by light stimulation of the contralateral eye. The chicks were adapted to the dark for 30 min prior to stimulation. Increase in $[K^+]_e$ was evoked by 20 light flashes at frequency 2 Hz (*light flashes*), by a single light flash (*flash*), and by application of continuous light for 5 min (*light*). Data from three chicks. (From Syková et al. 1990)

or introducing a drop of the taste-aversive substance methyl anthranilate into their beak (Fig. 26B; Jendelová et al. 1988; Syková et al. 1990).

4.2.2 *Dynamic $[K^+]_e$ Changes in the Mesencephalic Reticular Formation*

The MRF is recognized to be a structure with a high spontaneous activity level. The spontaneous activity of the neurones in the MRF changes in the course of sleep cycles and wakefulness (Kasamatsu 1970; Manohar et al. 1972), during barbiturate anaesthesia (Noda and Adey 1973), and in correlation to the heart and respiration rate (Sýkorová-Syková et al. 1968; Radil-Weiss and Syková 1977a,b). Numerous studies have shown that a great many neurones in the MRF display spontaneous activity characterized by an incidence of bursts of action potentials, i.e., of groups of a few action potentials following one another at shorter intervals than the intervals between the discharges (MacGregor et al. 1973; Škvařil et al. 1971; Verzeano and Calma 1954). The raised action potential frequency in the bursts may appear synchronously in neighbouring neurones of the MRF and is usually accompanied by an incidence of EEG spindles in the cerebral cortex (Syka et al. 1975).

Dynamic $[K^+]_e$ changes in the immediate vicinity of individual neurones in the MRF depend on the distance of the K^+-ISM from the neuronal membrane and on the frequency of spontaneous activity in that neurone (Syková et al. 1974). A 0.02–0.2 mmol/l increase in $[K^+]_e$ was observed near neurones which displayed spontaneous activity. During a burst of action potentials (Fig. 29A–F), $[K^+]_e$ steadily increases and continues to do so for a further 100–300 ms. The return of $[K^+]_e$ to the pre-burst value is usually followed by a new cycle of increased action potential frequency, which is again accompanied by an increase in $[K^+]_e$. When the averaging technique was used, changes in $[K^+]_e$ could be recorded after a single action potential. The size of the $[K^+]_e$ increment is proportional to the distance of the microelectrode from the neuronal membrane (Fig. 29G–H). The maximum $[K^+]_e$ change which was recorded after a single action potential appeared with a latency of 5–20 ms from termination of the action potential, and the return to resting level took about 60 ms (Syková et al. 1974). The observed changes amounted to only 0.01–0.02 mmol/l, but owing to the variations in resting $[K^+]_e$ activity which are associated with spontaneous activity, and since the accuracy of the calibration of microelectrodes is limited in the presence of a low K^+ concentration, these changes can only be approximately determined. The smaller the distance of the microelectrode from the neurone, the greater the recorded extracellular action potential, the greater the change in $[K^+]_e$, and the sooner the maximum change in $[K^+]_e$ is reached. Figure 2 shows that the possibility of bringing the microelectrode right up to the neurones or into the intercellular space is limited by the size of the microelectrode. $[K^+]_e$ changes in the narrow intercellular cleft after a single action potential are, therefore, in actual fact several times greater than the values we measure. This fact is supported by the studies of Frankenhauser and Hodgkin (1956), who showed that the K^+ increase in

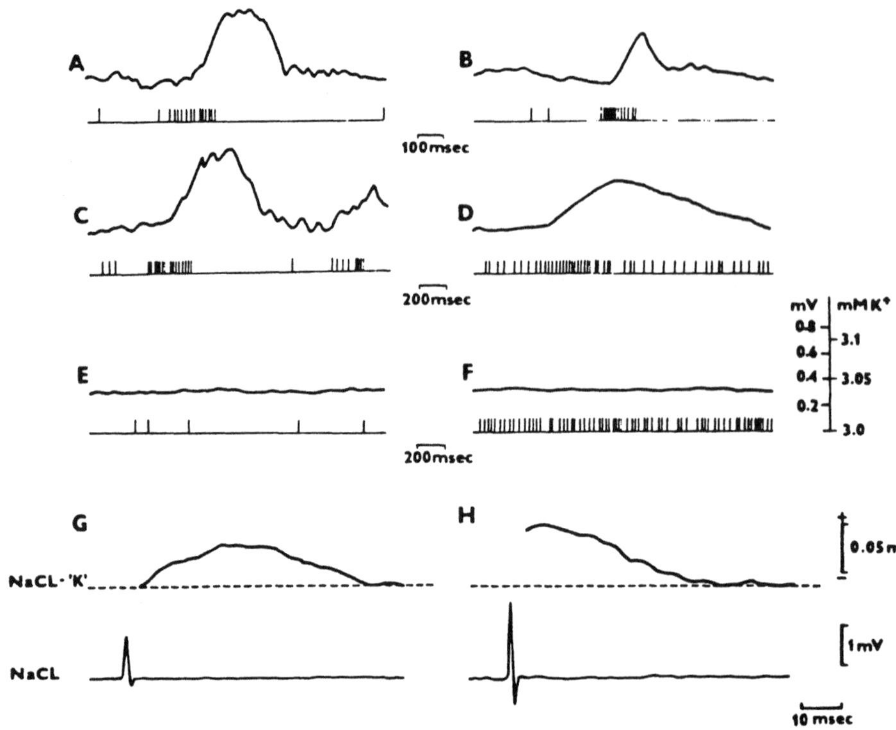

Fig. 29A–H. Elevation of $[K^+]_e$ associated with a spontaneous incidence of action potentials in the rat MRF. **A–F** Extracellular recordings of spontaneous neuronal activity (*bottom* recordings) accompanied by elevation of the $[K^+]_e$ (*top* recordings) in the vicinity of six neurones. The neuronal action potential bursts in **A–D** are accompanied by an increase in $[K^+]_e$; no such increase occurs in the vicinity of neurones with regular activity (**E**,**F**). **G**,**H** elevation of $[K^+]_e$ after completion of one action potential. The closer the electrode to the neurone, the larger the amplitude of the action potential (*NaCl*) and the greater the size and rate of the increase in $[K^+]_e$ (*NaCl-"K"*). The curves are averaged recordings from 128 samples (**G**) and 66 samples (**H**). (From Syková et al. 1974)

the intercellular space of squid giant axones after a single action potential amounted to 1.6 mmol/l.

Neuronal activity in the MRF is manifestly accompanied by the fluctuation of extracellular K^+. The dynamic changes evoked by the activity of given groups of neurones also influences adjacent neurones and glial cells in the vicinity, thereby helping to mediate interaction and synchronization of the activity of neuronal populations.

4.2.3 Dynamic $[K^+]_e$ Changes in the Cerebellum and Hippocampus

An increase in $[K^+]_e$ associated with activation of individual Purkinje cells via climbing fibres was found in the cat and rat cerebellum. The changes ranged from 0.3 to 0.5 mmol/l and their time course corresponded to that of evoked extracellular potentials (Bruggencate et al. 1976). The activation of Purkinje cells via climbing fibres exhibited rhythmic frequency changes which were associated with similar changes in olivary transmission. These slow rhythmic variations mirrored changes in $[K^+]_e$ and $[Ca^{2+}]_e$ (Stöckle and Bruggencate 1978, 1980). Stöckle and Bruggencate suggested that the rhythmic changes observed in extracellular ionic activities might affect excitability and transmission from parallel fibres to the Purkinje cells and thus help modulate the transmission of information in the cerebellar cortex.

An increase in $[K^+]_e$ of up to 8 mmol/l was observed during direct stimulation of the cerebellar surface (Nicholson et al. 1977, 1978). A 0.4 mmol/l decrease in $[Ca^{2+}]_e$ was simultaneously recorded. Laminar analysis showed that ion changes were greatest at the level of maximum stimulation of parallel fibres on the dendrites of Purkinje cells. The time course for these changes and pattern of redistribution resembled that observed in the spinal cord and cerebral cortex. Nicholson et al. (1978) demonstrated that a large proportion of K^+ (about 50%) came directly from parallel fibres. A $[K^+]_e$ increment of over 10 mmol/l was induced in cats by stimulation with a single electric pulse during the administration of aminopyridine or barium (Nicholson et al. 1976). The effect of ouabain on $[K^+]_e$ was the same as that seen in the spinal cord (see Fig. 5).

Several studies have demonstrated that $[K^+]_e$ changes in the cerebellum can influence the excitability of presynaptic afferent parallel fibres, even under physiological conditions (Kocsis et al. 1981, 1983; Malenka et al. 1981). Marked changes in excitability were evoked by electrical stimulation of parallel fibres with only ten impulses with a frequency of 50 Hz. Generally granule cells discharge in bursts of action potentials at a frequency of about 500 Hz (Eccles et al. 1966). In their experiments, Kocsis et al. (1983) demonstrated that small increases in $[K^+]_e$ may elicit an increase in the excitability of parallel fibres, whereas greater increases lead to a decrease in their excitability. As seen in the dorsal spinal horns, $[K^+]_e$ changes in the vicinity of the Purkinje cells and fibres in the cerebellum may function as an important mechanism which modifies the excitability of afferent fibres and postsynaptic elements (Malenka and Kocsis 1982; see Sect. 4.3).

The notion that K^+ accumulation in the ECS of the hippocampus as a result of neuronal activity could directly modify the neuronal response is not new (Hughes 1958; Green 1964). The use of ISMs has since confirmed $[K^+]_e$ and $[Ca^{2+}]_e$ changes in the rat hippocampus during electrical stimulation of the fimbria. Maximum increases (up to 12 mmol/l) were found in the pyramidal cell layer, where the simultaneously recorded field potential was the most positive (Krnjević et al. 1980, 1982b; Morris 1981). The increase in $[K^+]_e$ had a characteristic time course, a typical ceiling level, and a K^+ undershoot – as did the transient changes observed in the spinal cord and other parts of the brain. A series of further studies analyzing ionic changes, the mechanism of their origin, and their significance were carried out on hippocampal slices. The changes observed were similar in character to those seen as a result of in vivo experiments; the greatest changes were found in the pyramidal layer during orthodromic stimulation (Fritz and Gardner-Medwin 1976; Benninger et al. 1980). $[K^+]_e$ and $[Ca^{2+}]_e$ changes were closely interconnected and had the same depth profile (see Chap. 5). Prince et al. (1981) suggested that Ca^{2+}-activated K^+ permeability and synaptic processes were more relevant to the development of $[K^+]_e$ changes than action potentials in the axones and postsynaptic elements. In studies performed in the hippocampus in vivo and in vitro, they – and other authors – demonstrated that $[K^+]_e$ changes play an important role in the regulation of neuronal excitability and could act at both presynaptic and postsynaptic level (Poolos et al. 1987; Alger and Teyler 1978; Benninger et al. 1980; Hablitz and Lundervold 1981).

4.3 Functional Significance of $[K^+]_e$ Changes in the CNS

The functional significance of dynamic changes in K^+ concentration has been studied intensively over the past 18 years.
The most important findings relate to the role and influence of transient K^+ changes in:

- the depolarization of primary afferents and the mechanism of presynaptic inhibition
- the depolarization of postsynaptic elements and glial cells
- the release of transmitters
- the effectiveness of synaptic transmission
- the integration and synchronization of neuronal activity
- the origin of slow potentials
- the activation of membrane permeability for Ca^{2+} and inactivation of permeability for Na^+
- the effect on the Na^+/K^+ pump
- plastic changes in the CNS
- regeneration processes
- the development of the CNS
- pathophysiological processes

One aspect which often remains unclear is the causal relationship between the accumulation of K^+ and the process during which accumulation takes place, i.e., the question as to whether changes in $[K^+]_e$ are not just a concomitant phenomenon. However, dynamic changes in $[K^+]_e$ undoubtedly have a significant role in all excitable tissues and must be taken into account as a possible signalization, modulation, and feedback factor.

4.3.1 Role of K^+ in Presynaptic Inhibition

The efficacy of impulse transmission in the CNS is controlled by two types of inhibitory processes – postsynaptic and presynaptic inhibition. Postsynaptic inhibition is mediated by inhibitory synapses at which a specific transmitter (e.g., GABA, glycine) is released from synaptic terminals (Eccles 1964). During the activation of inhibitory synapses, permeability of the postsynaptic membrane for K^+ and Cl^- increases. These ionic fluxes tend to hold the membrane potential near its resting potential, or to slightly hyperpolarize the membrane, i.e., they work against the initiation of an action potential. The hyperpolarization responses which can be recorded with microelectrodes are therefore termed inhibitory postsynaptic potentials (IPSP).

Presynaptic inhibition is a less specific mechanism. The excitatory responses (e.g., EPSPs) are depressed without any change in the excitability of the postsynaptic membrane (for a review see Schmidt 1971; Levy 1977). Inhibition occurs in the region of primary afferent terminals, which are depolarized by a mechanism termed "primary afferent depolarization" (PAD), which reduces the size of action potentials travelling along already depolarized terminals. It has been demonstrated, in several types of synapses, that the higher (i.e., the more negative) the membrane potential of a nerve terminal, the greater the amount of transmitter released from that terminal (Hagiwara and Tasaki 1958; Takeuchi and Takeuchi 1962; Hubbard and Willis 1962; Kusano et al. 1967; Erulkar and Weight 1977). The correlations found between the time course of presynaptic inhibition and PAD and parallel alterations of both phenomena following physiological and pathophysiological manipulations (Schmidt 1971), suggest a causal relationship.

The mechanism of presynaptic inhibition was first described – and is still best understood – at spinal level (Frank and Fuortes 1957; Eccles et al. 1961, 1962; Eccles 1964), although it also exists in the brain. Monosynaptic EPSPs of spinal motoneurones exhibit a prolonged depression which is associated with PAD. PAD was first recorded by Gasser and Graham (1933) as a slow positive surface potential, i.e., as a cord dorsum potential (P wave). Later, it was recorded as a negative dorsal root potential (DRP) (Barron and Matthews 1938; Eccles and Malcolm 1946; Lloyd and McIntyre 1949). Barron and Matthews (1938) already suggested that DRPs could be attributed to changes in the ion concentration in the synaptic clefts around active neurones and non-myelinated endings of primary afferent fibres, i.e., by a mechanism analogous to the one responsible for development of the hyperpolarization phase following action potentials in peripheral nerves. Gray (1962, 1963), however, described axo-axonic synapses in the spinal cord and on the basis of this discovery Eccles et al. (1962) put forward a new theory of presynaptic inhibition. They postulated that PAD and reduced transmitter release is the outcome of activation of the axo-axonic synapses localized on primary afferent terminals. This activation of axo-axonic synapses via polysynaptic interneuronal pathways results in PAD and hence in a reduction in the amplitude of action potentials conducted by primary afferents. Eccles et al. (1963a) were also the first to suggest that the transmitter at axo-axonic synapses might be GABA; since then, their findings have been confirmed by a series of neuroanatomical, neurobiochemical, and neuropharmacological studies (for a review see Levy 1977).

It was not until several decades later, when measurement with ISMs showed that K^+ accumulated in specifically defined areas in the deep layers of the dorsal horns, that the ionic theory of Barron and Matthews (1938) was revived. The increase in K^+ concentration resulting from activation of neurones in the dorsal horns, non-myelinated endings of primary afferents, and non-myelinated high-threshold fibres (see Sect. 4.1) apparently depolarizes presynaptic endings and thus has practically the same effect as presynaptic axo-axonic synapses.

4.3.1.1 Depolarization of Primary Afferents

There is evidence to indicate that K^+ is involved in the generation of PAD, including the discovery that the maximum $[K^+]_e$ increment in the spinal cord occurs in the dorsal horn layers where most of the primary afferents terminate, i.e., there is evidence of an anatomical correlation between the localization of primary afferent terminals and the intraspinal localization of PAD (see Levy 1977). In the cat and rat, the maximum $[K^+]_e$ increase after stimulation of an afferent input was found in the deep layers of the dorsal horns, in Rexed's laminae III–VII (Kříž et al. 1974; Bruggencate et al. 1974; Krnjević and Morris 1974; Lothman and Somjen 1975). These are areas where primary afferent fibres terminate, where the maximum focal potentials corresponding to total activity of the spinal interneurones are to be found, and where PAD – tested by means of excitability changes in these fibres – is particularly intensive (Eccles et al. 1963b,c). Similar results have also been obtained in amphibians (Syková et al. 1976; Czéh et al. 1981). In the case of supramaximal electrical stimulation of low-threshold dorsal root fibres or peripheral nerve afferents, the maximum $[K^+]_e$ changes were localized in the deep layers of the dorsal horn, although when only skin nociceptors were stimulated, the maximum increase was found in the more superficial layers – in the area of the substantia gelatinosa Rolandi (see Fig. 17). These findings concur with anatomical evidence that cutaneous primary afferents terminate in the substantia gelatinosa Rolandi (Scheibel and Scheibel 1968), and agree with the findings of Wall and his colleagues (Howland et al. 1955; Wall 1962, 1964) which show that the main site of the depolarization of primary afferents which conduct nociceptive information in mammals is the substantia gelatinosa (laminae II–III, Rexed 1954).

These findings support the hypothesis that accumulated K^+ plays a role in the mechanism of PAD, but they do not demonstrate a causal relationship. As described in Sect. 3.2, interneurones are the main source of K^+ accumulation during stimulation of afferent input; primary afferent terminals and non-myelinated fibres do, however, provide a secondary source. Afferent fibres branch and their branches clearly overlap (Scheibel and Scheibel 1968; Székely 1976), such an anatomical organization making it is easy for K^+ released from activated primary afferent fibres to depolarize other primary afferents. A blockade of conduction in primary afferents has, in fact, been demonstrated in several studies in mammals. It is known that impulses entering the spinal cord

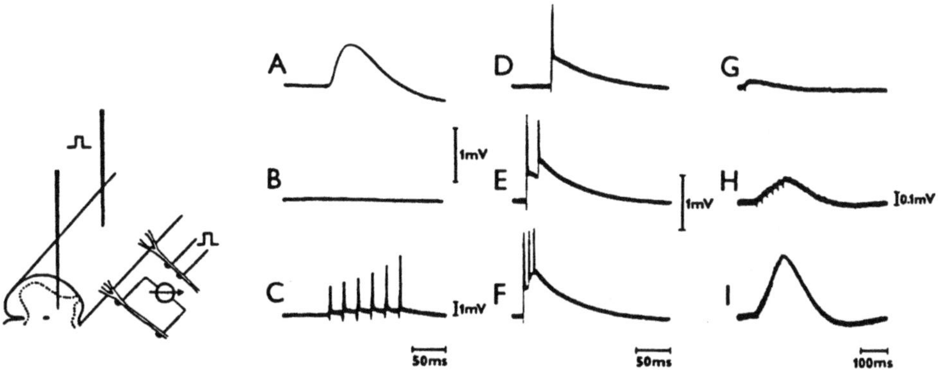

Fig. 30A–I. Effect of the administration of 20 mmol/l $MgSO_4$ on dorsal root potentials (*DRP*) and on the excitability of primary afferent fibres in the isolated frog spinal cord. *Left*, diagram of DRP recording after stimulation of the adjacent dorsal root or the intermediate region of the spinal cord with a pair of tungsten microelectrodes. **A** Control DRP. **B** Suppression of DRP by Mg^{2+}. **C** Antidromic action potentials recorded from a dorsal root after applying six pulses with tungsten microelectrodes. Note the increased excitability of primary afferent fibres. **D–F** Antidromic action potentials are accompanied by negative slow after-potentials which, if a few stimulating pulses are applied, display non-linear summation. **G–I** DRPs evoked by direct stimulation of the spinal cord display non-linear summation. (From Vyklický et al. 1976)

via dorsal roots and impulses travelling antidromically (dorsal root reflexes) can be blocked by previous activity in afferent fibres (Barron and Matthews 1935; Koketsu 1956a,b; Wall 1958). Grossman et al. (1973) observed that a blockade at branch points of the crustacean motoneurone axon by high-frequency stimulation was associated with an increase in the size of after-hyperpolarization of the action potential, testifying that K^+ accumulates in the periaxonal space. Such accumulation of K^+ can lead to a decrease in the amplitude of the next approaching action potential, owing to depolarization of the axonal membrane, a decrease in its impedance, and partial inactivation of sodium channels.

The notion that K^+ released from activated primary afferent fibres depolarizes other primary afferents in the spinal cord was confirmed in experiments which demonstrated increased excitability of primary afferents (Fig. 30) and "asynaptic" DRPs (Fig. 31) during a Mg^{2+} or Mn^{2+} blockade of synaptic activity (Vyklický et al. 1976; Syková and Vyklický 1977a; Nicoll 1979; Davidoff et al. 1988). Mg^{2+} blocked focal potentials and ventral root discharge which had been evoked by electrical stimulation of afferent input, but DRPs were not completely suppressed. The accumulation of K^+ in the ECS was likewise not completely suppressed. In our experiments, both DRPs and K^+ accumulation were depressed down to 10%–15% (Fig. 31), although in recent experiments on the isolated frog spinal cord in Ringer solution containing a physiological Mg^{2+} concentration (about 1 mmol/l) Davidoff et al. (1988) demonstrated depression to only 35%. These "residual" or asynaptic DRPs have their origin

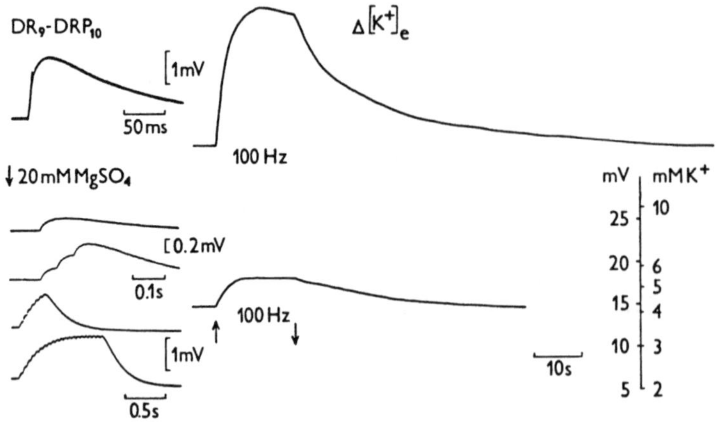

Fig. 31. Effect of blocking synaptic transmission on dorsal root potentials (*DR-DRP*) and on transient changes in $[K^+]_e$ induced in the isolated spinal cord of the frog by tetanic dorsal root stimulation at 100 Hz. *Left*, control DRPs and their depression after the 20 mmol/l $MgSO_4$ was added to Ringer solution. Note the non-linear summation of the DRPs under the effects of Mg^{2+} during repetitive stimulation; *right*, the increase in $[K^+]_e$ after tetanic stimulation (*above*) and the decrease after administering Mg^{2+} (*below*). (From Syková and Vyklický 1977a)

in depolarization of the fibres by K^+ which has been released directly from activated primary afferent fibres. Asynaptic DRPs were also found in the isolated frog spinal cord perfused with a Ca^{2+}–free Ringer solution containing Mg^{2+} (Syková and Vyklický 1978). During repetitive stimulation, asynaptic DRPs displayed a non-linear summation (Figs. 30 and 31; Syková and Vyklický 1977a; Nicoll 1979) which was similar to the summation of depolarizing responses found after repetitive stimulation in amphibian (Kuffler et al. 1966; Orkand et al. 1966) and mammalian (Takahashi and Tsuruhara 1987) glial cells. It is therefore reasonable to assume that K^+ released from the afferent fibres themselves depolarizes other afferent fibres, just as the K^+ released from activated neurones during intact synaptic activity depolarizes neighbouring neurones and primary afferents.

Further experiments have demonstrated the changes in excitability of primary afferent fibres and in DRP amplitude during perfusion of the spinal cord with a Ringer solution which had a raised K^+ concentration. It has been shown that perfusion with 10 mmol/l K^+ leads to an increase in $[K^+]_e$ in the dorsal horns, with simultaneous increased excitability of primary afferent endings, and depression of DRPs (Mellerová and Syková 1978; Krnjević and Morris 1981).

4.3.1.2 Effect of Picrotoxin and Bicuculline

The notion that depolarization of primary afferents partly results from the action of K^+ released from afferent fibres is supported by the finding that

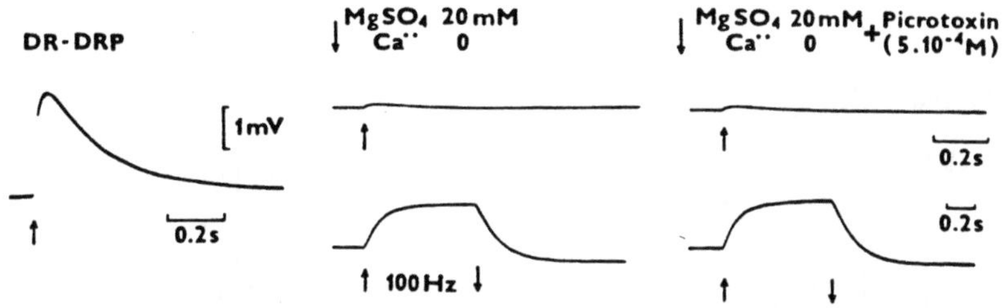

Fig. 32. Effect of picrotoxin on "non-synaptic" dorsal root potentials (*DR-DRP*) in the isolated frog spinal cord. *Left*, DR-DRP in normal Ringer solution; *middle*, effect of the administration of 20 mmol/l $MgSO_4$ in calcium-free Ringer solution on a non-synaptic DR-DRP evoked by applying one pulse or a series of pulses at a frequency of 100 Hz to the adjacent dorsal root; *right*, the administration of picrotoxin ($5 \cdot 10^{-4}$ mol/l) has no effect on the non-synaptic DR-DRP. (From Syková and Vyklický 1978)

asynaptic DRPs are resistant to the GABA antagonist picrotoxin (Fig. 32). Simultaneous recording of both $[K^+]_e$ changes and depolarization of primary afferent fibres in the frog spinal cord during application of picrotoxin provided evidence that the depolarization of primary afferent fibres has a dual mechanism.

The fast component of the DRP is severely inhibited by picrotoxin or bicuculline (Barker et al. 1975; Syková and Vyklický 1978; Nicoll 1979; Davidoff et al. 1980) and is thus largely the result of activation of axo-axonic synapses at primary afferent terminals. The slow (or late) DRP component is, however, enhanced and lasts longer after application of picrotoxin; postsynaptic neuronal activity is stimulated, and there are increased changes in $[K^+]_e$ (Fig. 33; Syková and Vyklický 1978; Syková 1981). Administration of picrotoxin led to a rise in $[K^+]_e$ in the dorsal horns of 1–4 mmol/l after peripheral nerve stimulation with only a single electric pulse. Spontaneous increases in $[K^+]_e$ occurred frequently and were accompanied by spontaneous DRPs (Syková and Vyklický 1978).

These findings testify to the fact that PAD has a dual mechanism, and the original objection that changes in $[K^+]_e$ and depolarization of the fibres have different time courses is no longer significant. We can thus conclude that the first fast component with a constant latency largely results from the action of a specific transmitter – GABA. The descending phase of this component is, however, inconstant and is often influenced by the functional state of the experimental animals. This phase is also much slower than it would be if it were attributable to the redistribution of released transmitters, and its time course is not exponential in character. The descending phase must, therefore, be influenced by some other mechanism which depolarizes primary afferent fibres over a longer period. The signals measured with K^+-ISMs and the aforementioned evidence suggests the involvement of K^+. It is possible, however, that the accumulation of K^+ also contributes to some extent to the first, faster phase of depolarization, since the K^+ signals picked up by K^+-ISMs are markedly delayed owing to the distance of the electrode from the membrane

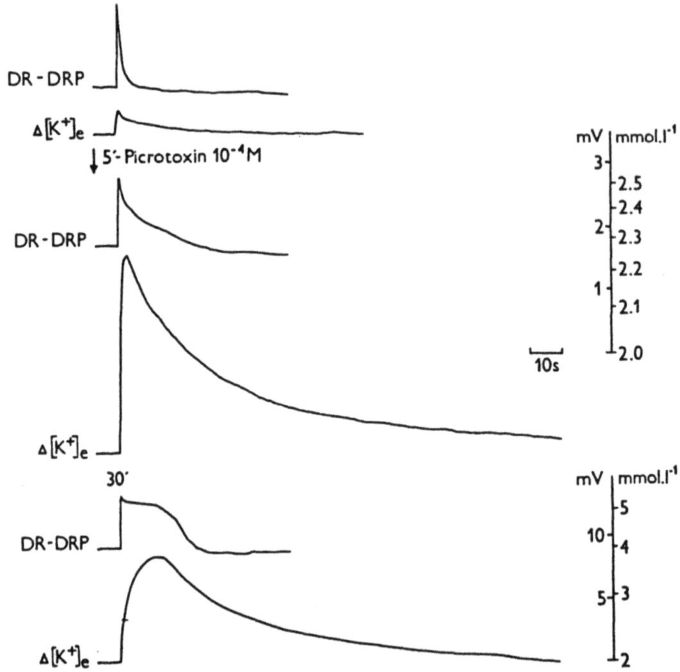

Fig. 33. Effect of picrotoxin (10^{-4} mol/l) on dorsal root potentials (*DR-DRP*) and on changes in $[K^+]_e$ in the isolated frog spinal cord. Note the prolongation of DRPs at 5 s and 30 s after application of picrotoxin and corresponding increase in $[K^+]_e$. (From Syková 1981)

source of K^+. The tip of the microelectrode (2–3 μm) cannot be localised as close to the membrane as the terminals of afferent fibres, which are separated from the membrane by a narrow synaptic cleft only 10–25 nm across (see Fig. 2).

As previously mentioned, after stimulation with a single electric pulse the synaptic component is larger than the potassium component. The K^+ component does, however, grow as electrical stimulation is repeated. Similarly, the second, longer K^+ component is also more pronounced after adequate stimulation of cutaneous nociceptors (Czéh et al. 1981). The presence of this second component during nociceptive stimulation of the hind limb with hot water (see Sect. 4.3.1) is clearly discernible in the descending phase of the first fast depolarization wave and attains up to 90% of the amplitude of the synaptic component. Changes in $[K^+]_e$ after nociceptive stimulation can likewise be compared with those after repetitive electrical stimulation rather than after stimulation by a single electric pulse. A single electric pulse typically evokes only 1–5 action potentials in spinal interneurones, whereas they respond to a nociceptive stimulus, such as the application of hot water or pinching, by high frequency firing which lasts several dozen seconds longer than the stimulation (see Fig. 19). It can thus be inferred that the late, protracted component which is seen in the depolarization of primary afferents after stimulation of cutaneous nociceptors is also due to the accumulation of K^+ in the ECS.

The hypothetical dual mechanism model for the depolarization of primary afferents has also been tested in experiments in which K^+ depolarization was superimposed on GABA or glutamate depolarization of primary afferents and vice versa (see Levy 1980). K^+ depolarization was found to be the same irrespective of whether it was superimposed on GABA or on glutamate depolarization, but the depolarization caused by GABA or on glutamate after K^+ depolarization was smaller than without K^+ depolarization. It can thus be demonstrated that the K^+ component of PAD can follow the synaptic component without losing any of its effectiveness. During tetanic stimulation, as during the continuous administration of GABA, desensitization of GABA-induced responses can presumbly be expected (Nistri and Corradetti 1978; Holobut and Niechaj 1973). In conclusion, it can be claimed that K^+ depolarization plays a significant role in prolonged – but probably non-specific – depolarization of primary afferent fibres.

4.3.2 Effect of K^+ Accumulation on Synaptic Transmission

An increase in $[K^+]_e$ does not, of course, only cause depolarization of primary afferents and an increase in the excitability of their endings, i.e., presynaptic inhibition; it also increases the excitability of postsynaptic elements. Depolarization of neurones actually makes them more excitable an thereby facilitates impulse transmission. These two simultaneous effects of an increase in $[K^+]_e$, i.e., presynaptic and postsynaptic, undoubtedly counteract each other. In this way, neuronal depolarization which has been induced by K^+, especially in the low concentrations which occur during stimulation with physiological stimuli, can lead to an increase in neuronal excitability which may outweigh the presynaptic inhibitory effect. In addition, an increase in $[K^+]_e$ can modify synaptic activity by stimulating transmitter release. Krnjević and Morris (1976) showed that an inverse correlation exists between the effectiveness of synaptic transmission and the excitability of primary afferents. Moreover, the increase in $[K^+]_e$ may also affect synaptic transmission by stimulating spontaneous transmitter release. Studies have shown that an increase in $[K^+]_e$ produces an increase in the frequency of miniature end-plate potentials (m.e.p.p.), which is a function of the membrane potential of motor nerve terminals (Liley 1956). Since the frequency of m.e.p.p. can also be altered by presynaptic depolarization (Del Castilo and Katz 1954; Liley 1956), this effect was attributed to presynaptic depolarization. However, there is evidence that changes in $[K^+]_e$ can also stimulate transmitter release directly (Gage and Quastel 1965; Cooke and Quastel 1973). The existence of this "non-specific" effect of K^+, i.e., an increase in transmitter release, was deduced as a result of experiments at the rat neuromuscular junction. Cooke and Quastel (1973) proposed that the accumulation of K^+ resulting from the increase in K^+ permeability during EPSPs led to a "specific" effect on transmitter release and possibly acted as a

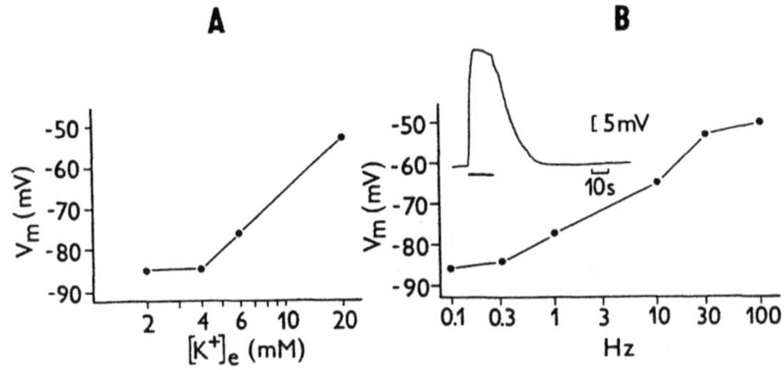

Fig. 34A,B. Effect of $[K^+]_e$ changes in the Ringer solution and of repetitive dorsal root stimulation on the membrane potential of a motoneurone in the isolated frog spinal cord. **A** Steady-state membrane potential during perfusion of the spinal cord with 2, 4, 6, and 20 mmol/l K^+. **B** Membrane potential of the same motoneurone 0.5 s after a 15 s dorsal root tetanus at the frequencies indicated. All values were measured during a single penetration of the motoneurone. *Inset*, record obtained with tetanic stimulation of 100 Hz for 15 s. (From Syková and Orkand 1980)

mechanism for the modulation of transmitter release from nerve endings, depending on the membrane potential of the target cells.

Heterosynaptic facilitation is thus probably a common phenomenon in the CNS during the accumulation of K^+. However, it is still difficult to ascertain whether the effect of K^+ accumulation on impulse transmission in the CNS is the outcome of its action on a presynaptic or a postsynaptic site.

4.3.2.1 Effect of K^+ on Neuronal Membrane Potential

An increase in K^+ depolarizes not only the glial cells (see Sect. 4.3.3), but also the membranes of neurones and fibres. The first study showing a quantitative relationship between $[K^+]_e$ and the membrane potential of the squid giant axone was published by Curtis and Cole (1942). The same relationship was later observed in isolated parts of the nervous system. Huxley and Stämpfli (1951) demonstrated a correlation between the $[K^+]_e$ and membrane potential in the node of Ranvier of an isolated single myelinated nerve fibre. Further data have been obtained on the membrane potentials of neurones in the nervous system of the leech (Kuffler and Nicholls 1966; Kuffler 1967). In the vertebrate CNS, similar studies have been carried out in the isolated spinal cord of the frog or toad, after perfusion with Ringer solution which had a raised K^+ concentration (Matsuura 1969; Syková and Orkand 1980). Intracellular recordings from motoneurones showed the resting membrane potential to be linearly correlated to the logarithm of the K^+ concentration, with a slope of 44 mV for a tenfold concentration change when K^+ concentration was 8 mmol/l or higher (Fig. 34A; Syková and Orkand 1980). With lower K^+ concentrations the slope was flatter, although even then the membrane potential fell significantly when perfusion lasted long enough to allow a new level to be established in the ECS. Matsuura (1969) also demonstrated that the motoneurone membranes were hyperpolarized when the K^+ concentration was lowered below the resting $[K^+]_e$. If the K^+ concentration in the perfusion solution was raised, the membrane resistance of the motoneurones decreased. The increase in $[K^+]_e$ evoked by repetitive dorsal root stimulation had a comparable depolarizing effect on the membrane potential of motoneurones (Fig. 34B). Intracellular recordings from glial cells show that the glial cell membrane potential in the frog spinal cord is linearly correlated – as found in *Necturus* (Orkand et al. 1966) – to the logarithm of the K^+ concentration, even when K^+ concentration is low (Fig. 8; Syková and Orkand 1980).

4.3.2.2 Effect of K^+ on Synaptic Potentials and Spontaneous Activity

During perfusion of the isolated frog or toad spinal cord with solutions containing different K^+ concentrations the amplitude of motoneurone action potentials and EPSPs was found to be directly proportional to resting potential changes (Matsuura 1969; Syková and Orkand 1980). A similar correlation was found when an increase in $[K^+]_e$ was evoked by tetanic stimulation of afferent input (Syková and Orkand 1980). The buildup of depression of EPSPs and decrease in membrane potential always followed the same time course, as did the return to the control value after solution with a raised $[K^+]_e$ concentration had been washed out (Fig. 35) and the drop in $[K^+]_e$ increment after tetanic stimulation (Fig. 36). Motoneurone depolarization produced either by raising the $[K^+]_e$ in the perfusion solution, or by tetanic stimulation, affects the size and duration of action potentials evoked by antidromic, orthodromic, or direct stimulation. In some motoneurones, Matsuura (1969) described a decrease to an almost complete block of action potentials as a result of raising $[K^+]_e$ to only 6 mmol/l. An increase in $[K^+]_e$ also prolonged the latency of antidromic action potentials, indicating that conduction velocity along the axones had slowed down. Other motoneurones, on the other hand, displayed only a minor change in action potentials when $[K^+]_e$ had increased above 10–12 mmol/l. Since K^+ accumulates preferentially in the deeper layers of the dorsal horns and in the intermediate region during stimulation of afferent input, the motoneurones located more superficially in the ventral horns are exposed to greater $[K^+]_e$ changes than those lying deeper in the ventral horns. Syková and Orkand (1980) found a decrease in the amplitude of antidromically evoked action potentials, diminution of the overshoot, and an increase in duration of the action potential after raising the K^+ concentration in the perfusate to 10–20 mmol/l, and after tetanic stimulation of the dorsal roots at a frequency of 10–30 Hz; as a result $[K^+]_e$ rose by several mmol/l in the relevant parts of the spinal cord. Conversely, antidromic tetanic stimulation with the maximum frequency which the motoneurones are capable of following (10–12 Hz) does not lead either to measurable changes in the membrane potential, or to a decrease in the size of action potentials and EPSPs. This is fully consistent with

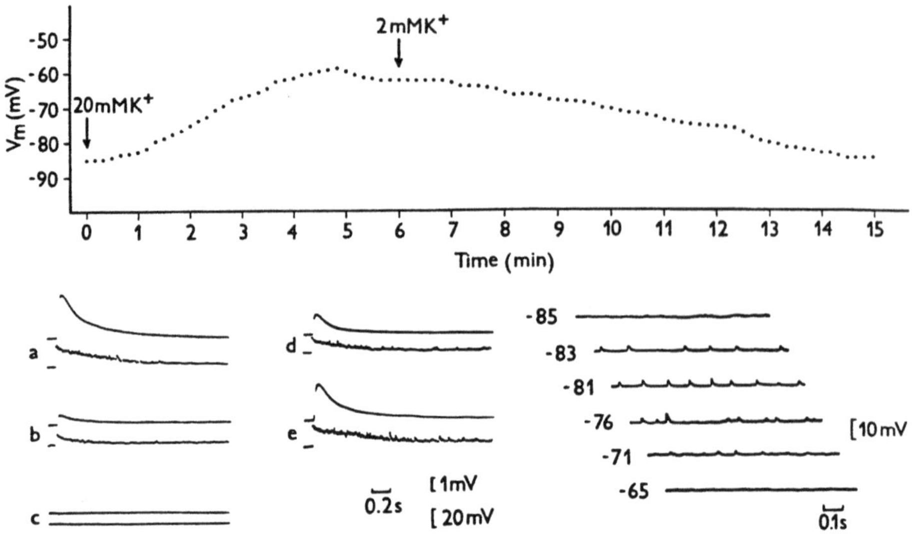

Fig. 35. Effect of elevated $[K^+]_e$ in the Ringer solution on motoneurone membrane potential (*above*), on DRPs, and on excitatory postsynaptic potentials (paired recordings *a–e*). *a*, control recordings; *b*, 2 min after adding 20 mmol/l K^+; *c*, 6 min after adding 20 mmol/l K^+; *d*, 4 min after adding 2 mmol/l K^+; *e*, 7 min after adding 2 mmol/l K^+. *Bottom right*, records of spontaneous synaptic activity intracellularly from motoneurone at indicated membrane potential. (From Syková and Orkand 1980)

the observation that in the spinal cord antidromic tetanic stimulation does not lead to an increase in $[K^+]_e$ greater than 0.2 mmol/l (Fig. 16; Kříž et al. 1974; Syková et al. 1976; Czéh et al. 1981; Syková and Orkand 1980).

An increase in $[K^+]_e$ in the perfusion solution or after tetanic stimulation is accompanied by depression or disappearance of EPSPs (Fig. 36). It is interesting to note that when tetanic stimulation or perfusion with a solution with a resting K^+ concentration is discontinued, recovery is accompanied by a period of enhancement and prolongation of EPSPs (Syková and Orkand 1980). Prolongation of EPSPs was observed during perfusion of the isolated frog spinal cord with a solution in which $[K^+]_e$ was raised by only 1–3 mmol/l (Fig. 37). Such small increases in $[K^+]_e$ are often associated with an increase in spontaneous activity. It is possible to observe spontaneous miniature EPSPs during intracellular recording from motoneurones (see Fig. 35). Excitability also often increases during small changes in $[K^+]_e$ which have been evoked by adequate stimulation of the skin. These results indicate that an increase in $[K^+]_e$ in the CNS has a dual effect on synaptic transmission. Small changes facilitate transmission in the CNS, while changes of over 6 mmol/l lead to its depression. Facilitation of transmission is evidently due to the postsynaptic action of K^+, since the amplitude of DRPs, which are one of the indicators of its presynaptic effect, decreases even when the $[K^+]_e$ increment is small (Mellerová and Syková 1978). Observations in the motoneurones undoubtedly apply to the interneurones in the dorsal horns and to other neurones in the

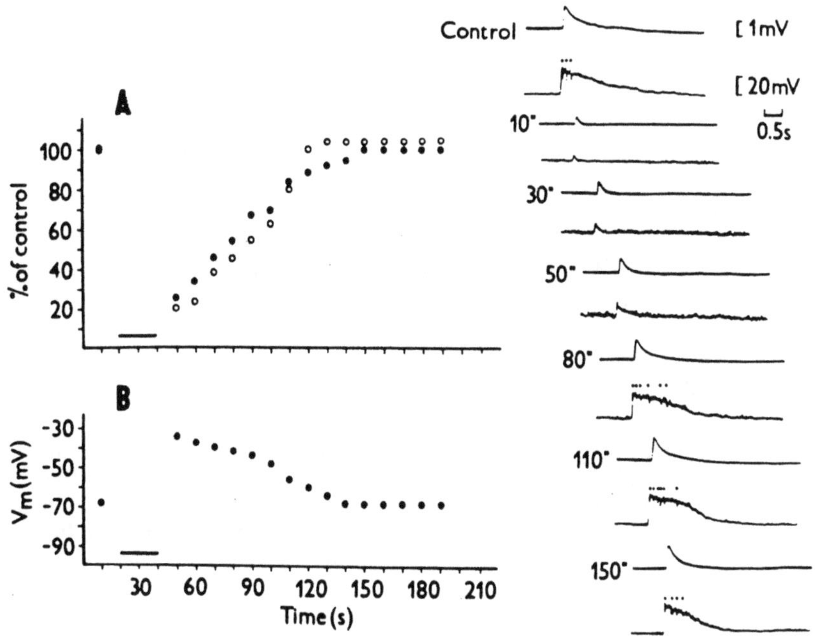

Fig. 36A,B. Depression of dorsal root potentials (DRP) and excitatory postsynaptic potentials (EPSP) in an isolated segment of the frog spinal cord after tetanic dorsal root stimulation. **A** Amplitudes of DR-DRP (*black circles*) and EPSP (*open circles*) after tetanic stimulation of the adjacent dorsal root at 100 Hz for 20 s. **B** Graph of motoneurone resting potential after same stimulation. The recordings on the *right* are pairs of DRP (*upper*) and EPSP (*lower*) before stimulation (*control*) and after tetanic stimulation. *Dots* above EPSP indicate occurrence of motoneurone spikes which are too brief to be visible at the slow sweep speed. Note the prolongation of the EPSP and increased numbers of spikes during recovery. (From Syková and Orkand 1980)

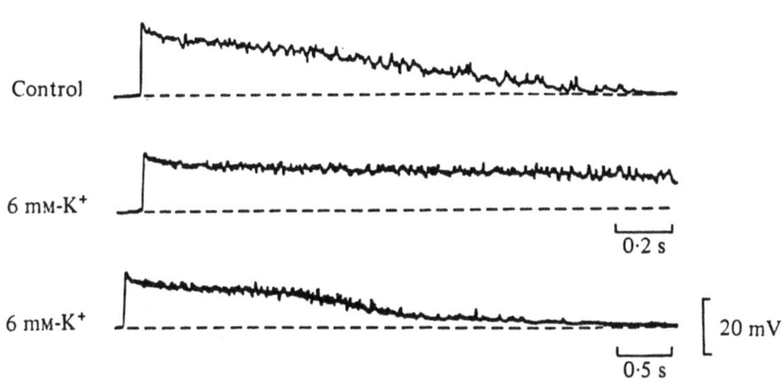

Fig. 37. Prolongation of motoneurone EPSPs in the isolated frog spinal cord during an increase of $[K^+]_e$ in perfusion fluid from 2 mmol/l K^+ (*control*) to 6 mmol/l. *Lowest trace* at 2.5 times slower sweep speed. (From Syková and Orkand 1980)

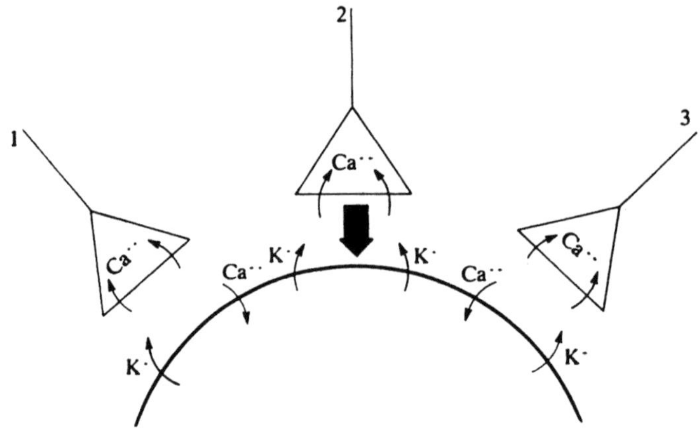

Fig. 38. Schematic diagram of a neurone (postsynaptic) with three separate convergent pathways (*1*, *2* and *3*). Repetitive firing of one excitatory pathway (*2*) would cause repetitive firing of the postsynaptic neurone, which would result in accumulation of K^+ in the ECS and a reduction in extracellular Ca^{2+}, which could in turn modulate transmission in the other synaptic pathways (*1* and *3*). (From Weight and Erulkar 1976)

CNS when they are faced with an increase in $[K^+]_e$ from 0.2 to 12.0 mmol/l. It can thus be concluded that excitability of neurones can either be significantly increased or depressed according to the intensity of the stimulation of afferent input.

The postsynaptic effect of K^+ accumulation, together with its presynaptic effect, represents a non-synaptic, negative-positive feedback mechanism for the regulation of impulse transmission in the CNS (Syková et al. 1974; Weight and Erulkar 1976; Syková and Orkand 1980; Kocsis et al. 1983; Poolos et al. 1987; Syková 1987).

Figure 38 illustrates one of the possible ways in which this feedback inhibition in the CNS could operate. The repeated neuronal discharges observed in the CNS during stimulation, and also spontaneously in association with physiological functions (Syková et al. 1974; Radil-Weiss and Syková 1977a,b), can be inhibited by the accumulation of K^+ in the ECS. K^+ which has been released from a neurone stimulated via excitatory pathways can cause feedback inhibition of transmitter release from the endings of other pathways which terminate on this and neighbouring neurones. As I have previously described, the elevation of $[K^+]_e$ during repetitive discharge can initially lead to the facilitation of transmitter release and to an increase in the frequency of action potentials – as in the case of action potential bursts. It is feasible that when a given K^+ level which depolarizes presynaptic axones has been reached, neuronal activity is fully inhibited. In this way, neuronal populations can defend themselves against harmful epileptiform activity, and ionic changes in the ECS act as an integrative and protective mechanism in the CNS (Syková 1983). Spira et al. (1984) showed that in insects (Periplaneta americana) the increase in $[K^+]_e$ during a single action potential leads to reciprocal synaptic interaction between

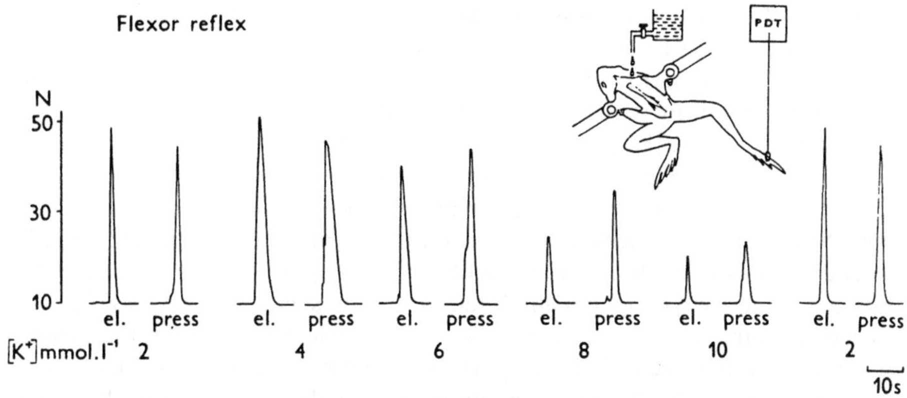

Fig. 39. Recordings of the frog's flexor reflex during superfusion of the spinal cord in vivo with Ringer solution containing 2 mmol/l K^+ (control) and 4, 6, 8, 10, and 2 mmol/l K^+. The flexor reflex was elicited by a short series of transcutaneously applied electric pulses (*el.*) or by pressing a toe of the foot with a forceps. Before any recordings were made, 30 min were allowed for superfusion with Ringer solution containing a higher concentration of K^+, and a further 60 min were allowed for recovery. The diagram illustrates the experimental set-up, in which the spinal cord was isolated from the spinal canal, except for dorsal and ventral roots IX and X. The spinal cord was superfused continuously with oxygenated Ringer solution. *PDT*, photoelectric displacement transducer. (From Vyklický and Syková 1980)

interneurones. These results were interpreted as evidence of the existence of physiological synaptic interaction assured by K^+.

4.3.2.3 *Effect of K^+ on Flexor Reflex*

The accumulation of K^+ during peripheral electrical stimulation, injury, and chronic pain can markedly affect spinal reflexes. In experiments on the frog spinal cord in situ, the cord was exposed and perfused with Ringer solution containing a raised K^+ concentration; the flexor reflex elicited by the application of a nociceptive stimulus (e.g., pinching or a short series of electric pulses applied through electrodes attached to the skin of a paw) was then studied (Fig. 39; Vyklický and Syková 1980). An increase in the K^+ concentration from 2 to 4 mmol/l prolonged and intensified the hind limb flexor reflex. If the spinal cord was perfused with solution in which the K^+ concentration was more than 6 mmol/l, the flexor reflex was shorter and weaker, i.e., contraction power was smaller. These results are again indicative of a dual effect of K^+ on impulse transmission in the spinal cord. An increase of 2–4 mmol/l in $[K^+]_e$, which can be produced by adequate stimulation or by injury, facilitates the flexor reflex. Greater increases in $[K^+]_e$, which can be caused by extensive injury, damage to the spinal cord itself, ischaemic processes in the spinal cord, or electrostimulation (including painless transcutaneous electrical stimulation) (Czéh et al. 1981; Syková et al. 1985b; Svoboda et al. 1988), can lead to inhibition of impulse transmission. There is an apparent discrepancy between the aforementioned results and the findings described by Baylor and Nicholls (1969) and Weight and Erulkar (1976), who observed an exclusively inhibitory effect as a result of increased $[K^+]_e$. These authors, however, studied the effect of solutions in which $[K^+]_e$ was raised by 4–9 mmol/l; our experiments demonstrated that an increase of only 2 mmol/l was found to have a facilitatory effect. Experiments on the isolated frog spinal cord connected to the isolated hind limb (Syková and Vyklický 1979) showed that ventral root potentials (VRPs) evoked by adequate stimulation (tactile or nociceptive) are enhanced in the presence of a low $[K^+]_e$ and inhibited in the presence of a high $[K^+]_e$, i.e. exceeding 6 mmol/l (Syková et al. 1985b). The exact site at which the facilitatory effects of increased $[K^+]_e$ on spinal cord transmission occur is still obscure. It is, however, likely that the excitability of secondary neurones in the dorsal horns increases together with spontaneous transmitter release, which would more than compensate for the deficit in synaptic transmission which results from relatively weak PAD.

4.3.3 K^+ Accumulation and Glial Cell Function

Orkand et al. (1966) and Syková and Orkand (1980) showed that glial cells are depolarized by K^+ which has been released from active neurones. The finding that the membrane resistance of glial cells falls only slightly during a depolarizing shift of the membrane potential concurs with the concept that depolarization is caused by an increase in $[K^+]_e$. Extracellular sustained potential (SP) shifts were found to have exactly the same time course as depolarization shifts in the glial cells of both the cerebral cortex (Castellucci and Goldring 1970; Ransom and Goldring 1973) and spinal cord (Somjen 1970). It has been suggested that the main source of SP shifts is the current flowing through the ECS as a result of the depolarization of glial cells; this depolarization has a similar time course to the changes which take place in $[K^+]_e$, and the sites are the same (Fig. 4A; Lothman and Somjen 1975; Lothman et al. 1975; Kříž et al. 1975; Vyklický et al. 1975). In regions with a high $[K^+]_e$, K^+ is taken up by the glial cells, whereas in places with a low $[K^+]_e$ it is released by them. There is thus a current flowing through the glial syncytium. In the ECS the reverse current is carried primarily by Na^+ and Cl^- and has a tendency to transport Na^+ back to the area of active neurones. This spatial buffer system has been described in Sect. 3.3.2.1. Glial cells thus act as a source of extracellular current although the function of such an extracellular current gradient is still uncertain. Possibly the current modifies the behaviour of the neurones (neuronal excitability and action potential frequency), either directly or by altering the polarization of molecules in the extracellular fluid. Neuronal excitability can also be modified by the increase in glial cell volume as a result of an increase in $[K^+]_e$; this will be discussed in Chap. 7.

The spatial buffer mechanism is a "space-dependent" form of K^+ redistribution (Fig. 3; Nicholson 1980). Space-independent processes ("net uptake") include (a) stimulation of Na^+/K^+ ATPase, (b) K^+, Cl^- or Na^+, K^+, Cl^- cotransport, (c) passive K^+, Cl^- uptake conforming to Donnan's equilibrium, and (d) K^+/H^+ exchange. All these mechanisms evidently play an important role in extracellular K^+ homeostasis.

Accumulation of K^+ in the ECS also influences the metabolism of the glial cells. Numerous experiments have shown that glucose, amino acids, and ions are exchanged between the glial cells and ECS and that this can modify neuronal function (Hertz 1979; Walz and Hertz 1983; Sears 1982; Treherne 1981; Walz 1989). It has been demonstrated with the aid of fluorimetric techniques that an increase in $[K^+]_e$ of 3–9 mmol/l reduces the pyridine nucleotides (NADH) level in glial cells (Orkand et al. 1973), and an increase of 12–15 mmol/l leads to greater accumulation of $[^{14}C]$ glucose in the glial cells (Salem et al. 1975; Cummins et al. 1979). An increase in $[K^+]_e$ also stimulates GABA release from the glial cells of rat dorsal ganglia (Minchin and Iversen 1974). Pentreath and Kai-Kai (1982) described raised incorporation of radiolabelled 2-deoxyglucose into the glycogen of glial cells in the vicinity of antidromically stimulated nerves. The incubation of ganglia in saline with a raised K^+ concentration had a similar effect; the maximum level was reached after an increase in $[K^+]_e$ of only 4 mmol/l, i.e., the value observed in healthy nervous tissue during stimulation of afferent input. The authors suggest that glucose uptake and glycogen synthesis in the glial cells rise during neuronal activity, in which case the glycogen reserves could act as a source of energy for neurones if the local blood glucose level is inadequate. A few enzymes have been found to be localized specifically in the astrocytes, such as glutamine synthetase, which forms glutamine from glutamate and ammonia – glutamine is also an important substrate for other reactions. Another important enzyme is carbonic anhydrase (see Chap. 6), which stimulates carbonic acid synthesis from CO_2. This reaction probably reduces the amount of CO_2 in the vicinity of the neurones (Pentreath 1982).

In conclusion, it must be admitted that the role of the glial cells, their relationship to K^+ homeostasis, and their functional significance for neuronal activity, pathological states, plastic changes, regeneration, etc. remain to be established. It is evident, however, that glial cells help keep the microenvironment of the neurones stable. The neurones, for their part, transmit signals to the glial cells in their vicinity by releasing K^+ into the intercellular spaces during activity. Glial cells can accept excess K^+ and thereby protect the neurones from prolonged changes in the extracellular K^+ concentration.

4.3.4 K^+ Accumulation and the Therapeutic Effect of Electrostimulation

It has repeatedly been demonstrated that direct or indirect stimulation of the spinal cord or stimulation of afferent fibres results in the relief of pain as well as beneficial muscle contraction in paretic and paralysed extremities (for reviews see Anderson 1975; Vodovnik 1981). Direct spinal cord stimulation (Campbell 1981), and the activation of peripheral nerve fibres by transcuta-

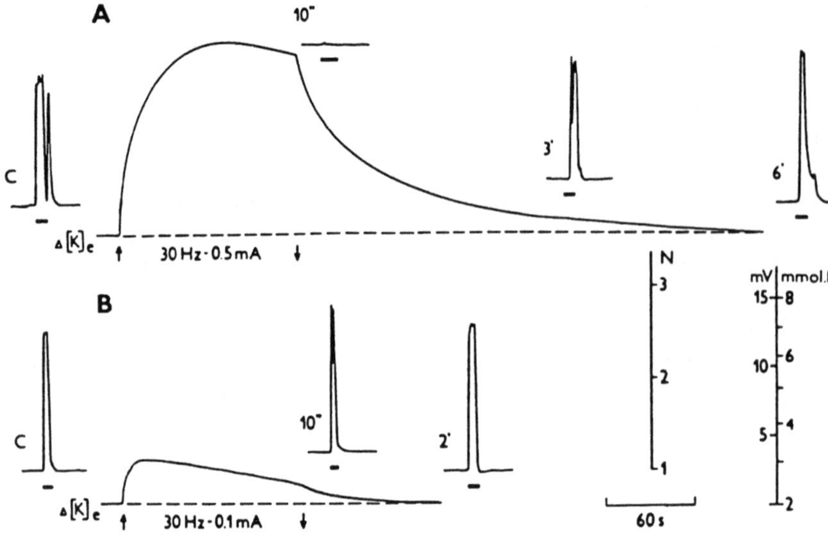

Fig. 40A,B. Elevation of $[K^+]_e$ in the frog spinal cord during repetitive electrical stimulation of the skin of a hind limb (30 Hz, 2 min) and simultaneous depression of the flexor reflex elicited by a nociceptive stimulus. The application of a drop of hot water (60°) to the skin of the paw was used as a nociceptive stimulus. The stimulation intensities were 0.5 mA (**A**) and 0.1 mA (**B**). The flexor reflex was recorded with a photoelectric displacement transducer 10 s before starting repetitive electrical stimulation (*inset C*), and 3 and 6 min (**A**) and 10 s and 2 min (**B**) after discontinuing it. (From Syková et al. 1985b)

neous electrostimulation or acupuncture are widely used for therapeutic purposes and for their analgesic effects in man and experimental animals (Melzack 1975; Long and Hagfors 1975; Hiedl et al. 1979; Woolf 1979). The final effect of stimulation far exceeds its local influence at the site of treatment. Many studies have dealt with the mechanism of the central effect of stimulation, but research is still at the hypothetical stage. One hypothesis which has recently received considerable support is the suggestion that the effects of stimulation (especially analgesic effects) are due to the release of endogenous opioid peptides (endorphin, dynorphin, and enkephalins) in either the brain or spinal cord (for reviews see Kosterlitz and McKnight 1981; Yaksh 1981; Nyberg et al. 1983; He 1987). Other studies confirm that the decrease in the pain threshold after repetitive electrical stimulation is not affected by naloxone and that it may be mediated by a non-opioid mechanism (Ernst et al. 1986).

In this connection, it should not be forgotten that stimulation of the CNS and peripheral nerves, including transcutaneous stimulation, leads to an increase in $[K^+]_e$ in the dorsal horns of both mammals and amphibians which could also contribute to the therapeutic effect of electrostimulation. In the same way that K^+ accumulation can have both a facilitatory and an inhibitory effect on spinal transmission (see Sects. 4.3.1 and 4.3.2), electrostimulation can also have a facilitatory or an inhibitory effect according to its frequency, intensity, and

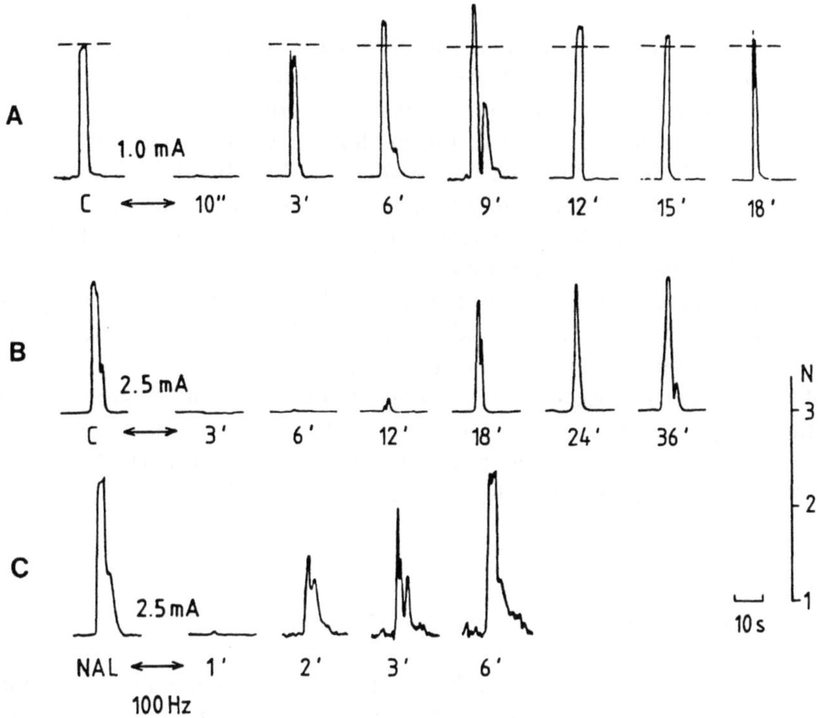

Fig. 41A–C. Effect of repetitive electrical stimulation of the skin (100 Hz, 1 min) on the frog flexor reflex elicited by a thermal nociceptive stimulus. **A** After stimulation with an intensity of 1.0 mA, the flexor reflex was at first depressed during the elevation of $[K^+]_e$ (see also Fig. 40), but this was followed by an increase which persisted for up to 20 min. **B** After stimulation with the higher intensity (2.5 mA), the flexor reflex was depressed for up to 35 min. **C** When naloxone (10^{-6} mmol/l) was applied to the surface of the spinal cord, long-term post-stimulation inhibition of the flexor reflex was suppressed, but short-term depression of the reflex corresponding to the increase in $[K^+]_e$ persisted. *C* and *NAL*, control recordings of the flexor reflex before stimulation. (From Syková 1987)

duration (Syková et al. 1985b). Figure 40 gives examples of inhibition of the flexor reflex elicited by a nociceptive stimulus (the application of a small amount of hot water to the skin of the frog's hind limb) following transcutaneous electrical stimulation of different intensities; the flexor reflex was, in fact, inhibited during the increase in $[K^+]_e$. The time course of the return of $[K^+]_e$ to the original, prestimulation level was the same as that for the return of the flexor reflex to normal values. This effect of electrostimulation on the flexor reflex persisted after the administration of naloxone (Fig. 41; Syková 1987). Transcutaneous stimulation, which raised $[K^+]_e$ by 2–6 mmol/1, further inhibited spontaneous activity, EPSPs, VRPs, and DRPs. The high-threshold myelinated A delta fibres and unmyelinated "nociceptive" C fibres are probably even more sensitive to the $[K^+]_e$ increase evoked by transcutaneous stimulation than low-threshold afferents.

Stimulation of A fibres has been shown to reduce the responses of second order cells in the spinal cord already evoked via C fibres (Mendell 1966; Brown et al. 1973; Cervero et al. 1976) by a mechanism possibly involving presynaptic and/or postsynaptic inhibition. Gating of noxious sensory input in the dorsal horn of the spinal cord due to the modulating effect of large fibre activity on that of small fibres (Melzack and Wall 1965; Wall 1978) may therefore be the first modulatory mechanism triggered by transcutaneous stimulation. Ionic changes – in particular an increase in $[K^+]_e$, which always results from activation of both neurones and primary afferents – are one of the likely explanations for the effects of stimulation at spinal cord level. Changes in $[K^+]_e$ can cause depolarization directly and indirectly by inducing greater release of opioid peptides from the endings of spinal primary afferents (Hájek and Syková 1981), or the release of other inhibitory and excitatory transmitters. One of the likely candidates is the potassium-induced release of GABA from axo-axonic synapses; this results in the depolarization of primary afferent fibres by altering their permeability for Cl^- (Nistri et al. 1974; Deschenes et al. 1976; Gallagher et al. 1978).

4.3.5 Other Functional Correlates of a $[K^+]_e$ Increase

An increase in $[K^+]_e$ may influence the affinity of membrane receptors for neurotransmitters. Ben-Ari et al. (1979) suggested that the decrease in the effectiveness of inhibitory transmission after tetanic stimulation of hippocampal pathways resulted in part from the reduced potency of the inhibitory transmitter GABA. A possible cause of "desensitization" to GABA is the change in $[K^+]_e$ and/or in $[Ca^{2+}]_e$ produced by repetitive stimulation (Krnjevič and Morris 1981). Tetanic stimulation of hippocampal pathways also causes prolonged depression of glutamate activity (Lynch et al. 1976). Increases of 1, 4, and 10 mmol/l in $[K^+]_e$ had statistically significant effects on the binding capacity of opiate receptors for ^{3}H-leucine enkephalin (LE) in the frog spinal cord (Hájek and Syková 1981). ^{3}H-LE binding increased significantly when $[K^+]_e$ in the isolated cord was raised from 2 to 3 mmol/l. Conversely, if the perfusion solution contained 6 or 12 mmol/l K^+, ^{3}H-LE binding decreased. The decrease in binding after exposure to 6 mmol/l can be attributed to a change in the oxidation-reduction state of the receptor proteins, whereas the changes evoked by a 12 mmol/l K^+ concentration were apparently due to structural changes in the opiate receptors themselves (Hájek and Syková 1981).

Kuschinsky and Wahl (1978) reported that a linear correlation exists between the diameter of the pial arteries and $[K^+]_e$ in the perivascular space – but only up to a K^+ concentration of 10 mmol/l. When $[K^+]_e$ was raised to 20 mmol/l there was no further increase in vascular diameter, and increases above 20 mmol/l were followed by vasoconstriction. Similar results were obtained in the hypothalamus (Cameron and Caronna 1976). An increase in $[K^+]_e$ to 11.9 mmol/l raises local glucose (2-deoxy-D-glucose) uptake and, together with an increase in the local blood flow, can lead to stimulation of metabolism and protein synthesis in active (stimulated) brain structures. In fact, a perivascular increase in K^+ or H^+ concentration, or a decrease in Ca^{2+} concentration all (individually or in combination) constitute a strong vasodilatatory stimulus.

Lipton and Heimbach (1977) found an increase in protein synthesis in guinea-pig hippocampal tissue slices perfused with a solution in which $[K^+]_e$ had been raised by 1.3–8.1 mmol/l. The incorporation of lysine into the proteins was directly proportional to the logarithm of $[K^+]_e$ and in the presence of the given concentrations it rose by up to 100% $[K^+]_e$ changes did not affect the tissue ATP or creatinine phosphate level, indicating that the observed effect on amino acid incorporation was not due to an effect on energy metabolism. These results provide a possible explanation of the way in which neuronal activity is coupled to changes in proteosynthesis and probably also to plastic changes in the CNS. They also support the hypothesis of Izquierdo et al. (1971) and Izquierdo (1972) that K^+ changes possibly contribute to the regulation of hippocampal function and the establishment of long-term memory. It is also probable that, owing to their regulatory effect on the composition of the microenvironment, glia – especially astrocytes – are involved in the modulation of neuronal responsiveness and thus behaviour. To date, research indicates that glial cells regulate extracellular K^+ and Ca^{2+} in addition to neurotransmitter glutamate and GABA – important candidates for the modulation of neuronal sensitivity (see Laming 1989). On intensive stimulation the number of glial satellite cells has been shown to increase in nerve centres (Alexandrovskaya and Chizhenova 1979; Sotnikov and Roitbak 1980). Roitbak also proposed that the basis of the conditioned reflex is the myelinization of axones in the cerebral cortex (Roitbak 1988)

Sussdorf and Campenot (1986) studied the growth of neurites in tissue cultures of sensory neurones, spinal neurones, and the neurones of sympathetic ganglia in medium containing different K^+ concentrations. They demonstrated that neurites were unable to grow from a compartment with a low K^+ concentration (5 mmol/l) to a compartment with a high K^+ concentration (20 mmol/l). Post-axotomy regeneration of the neurites of sensory neurones was also inhibited after an increase in $[K^+]_e$ (Sussdorf and Campenot 1986; Campenot 1984). It is indeed possible that the $[K^+]_e$ level plays and important role in the development of the CNS and in regeneration processes.

4.3.6 K^+ Accumulation and Its Functional Significance in Pathological Processes

The high density of neurones, fibres, and glial cells in the CNS (only about one quarter of which is taken up by the ECS) leads, in pathological situations, to intensive activity of neuronal elements and hence to enormous accumulation of K^+ in the ECS, accompanied by other ionic changes. The suppression of the metabolism of neuronal elements under pathological conditions may, nevertheless, also, cause an increase in $[K^+]_e$ owing to reduced efficiency of the Na^+/K^+ pump.

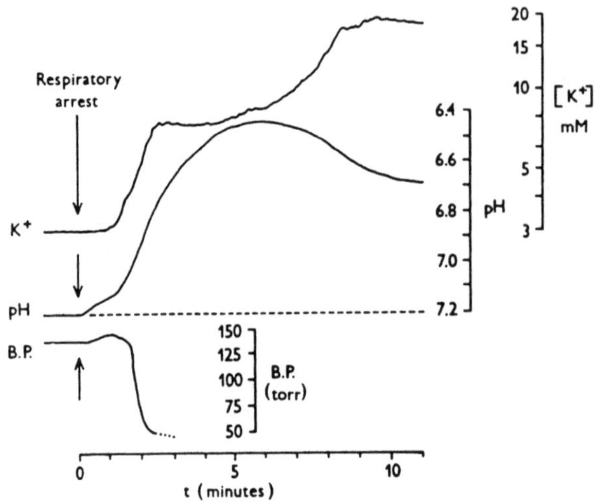

Fig. 42. Increase of $[K^+]_o$ and decrease of pH_e in the L_4 spinal segment at a depth of 600 μm as recorded after respiratory arrest. *B.P.*, concomitantly recorded changes in blood pressure. (From Syková and Svoboda 1990)

4.3.6.1 $[K^+]_e$ Changes During Ischaemia and Hypoxia

A great many authors have demonstrated K^+ accumulation in the ECS of the brain and spinal cord during hypoxia (Křivánek and Bureš 1960; Collewijn and Schade 1962; Meyer et al. 1962; Vyskočil et al. 1972; Prince et al. 1973; Morris 1974; Kirshner et al. 1976; Kříž et al. 1975; Hansen 1977; Syková and Svoboda 1990), ischaemia (Astrup et al. 1977, 1980; Branston et al. 1977; Bureš and Burešová 1981; Hansen and Zeuthen 1981), and hypoglycaemia (Astrup and Norberg 1976). These studies leave no doubt that the increase in $[K^+]_e$ has a biphasic course. Initially, there is a rise in $[K^+]_e$ from 3 mmol/l to about 10–12 mmol/l accompanied by small extracellular changes in the concentration of other ions with the exception of H^+; there is then an instantaneous shift to acidosis (Fig. 42). This $[K^+]_e$ increment is probably associated with intensive neuronal activity, but there is only partial impairment of the mechanisms responsible for the clearance (reabsorption) of K^+ which has accumulated in the ECS. The second, massive increase in $[K^+]_e$ to between 30 and 80 mmol/l results from an increase in cell membrane permeability. The abrupt $[K^+]_e$ increment is accompanied by a $[Na^+]_e$ and $[Cl^-]_e$ decrement (Hansen and Zeuthen 1981), and also by a rapid negative shift in the local electric potential, i.e., by anoxic depolarization. This latter phase is characterized by complete loss of K^+ clearance (Branston et al. 1977), evidently owing to ATP depletion and failure of the Na^+/K^+ pump. Hansen (1977) suggested that immature animals are more resistant to anoxia because they are more able to maintain near normal K^+ gradients across the cell membranes in the brain.

The kinetics of K^+ clearance from the ECS after ischaemia and anoxia are similar to those described after electrical simulation, i.e., they are exponential

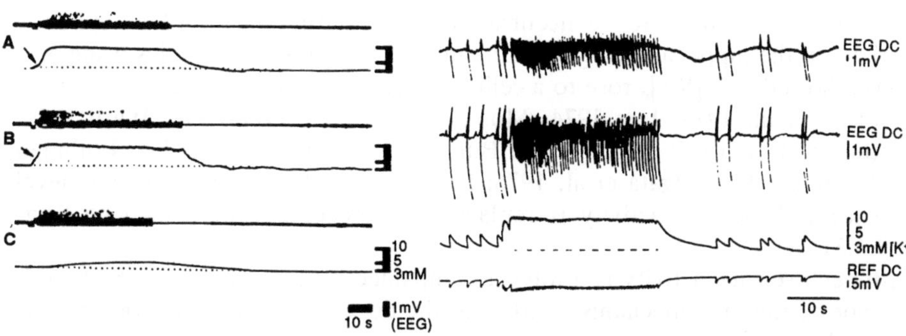

Fig. 43. Changes in K^+ activity during cortical seizures. *Left*, laminar analysis of $[K^+]_e$ responses during major propagated seizures in the cat cortex. *Top trace*, surface EEG; *bottom trace*, dotted lines at 3.1 mmol/l $[K^+]_e$. Depths: *A*, 300 μm; *B*, 1200 μm; *C*, 2400 μm. Standard supramaximal stimulus train parameters (100 V, 60 Hz, 2 s). (From Sypert and Ward 1974). *Right*, interictal and ictal events recorded from the cortical surface of the rabbit (*EEG DC* and *EEG DC*), and by K^+-selective (*third trace*), and reference (*REF DC*) microelectrodes located about 1 mm below the pial surface. The ictal period is marked by a *dashed line* below the third trace which also denotes the 3 mmol/l $[K^+]_e$ level. (From Futamachi et al. 1974)

in character, though much slower than those observed after electrical stimulation. Theoretically, K^+ should also be removed by capillary transport, diffusion in the ECS, and reabsorption into the cells. The permeability of the capillaries for K^+ is very slight, however (Hansen et al. 1977; Mutsuga et al. 1976), and there is nothing to suggest increased permeability of the blood-brain barrier during and after a short period of anoxia or ischaemia (Ito et al. 1976). Diffusion of K^+ in the ECS may also be limited owing to the absence of concentration gradients. The active reabsorption of K^+ by cellular elements is, therefore, probably the only important mechanism for the normalization of $[K^+]_e$ (Hansen et al. 1980). Blank and Kirshner (1977) showed that in cats resuscitated after hypoxia or anoxia, $[K^+]_e$ fell to a level which remained several mmol higher than that seen in normal cats; in some animals the $[K^+]_e$ level returned, very gradually, to normal, but in others it remained permanently raised. These findings indicate impairment of the Na^+/K^+ pump, the repair of which might require new RNA and protein synthesis.

4.3.6.2 K^+, Epilepsy, and Epileptiform Activity

Epilepsy is characterized by seizures with sensory and motor manifestations and abnormal neuronal activity. The seizures may differ as regards their aetiology, but they are all accompanied by an increase in $[K^+]_e$ associated with paroxysmal activity (see Fig. 43). Some authors actually suggest that the increase in $[K^+]_e$ may be a common cause of paroxysmal activity (Green 1964; Izquierdo and Izquierdo 1967; Fertzier and Ranck 1970; Pollen and Trachtenberg 1970; Dichter et al. 1972). Perfusion of the cerebral cortex with a solution with a high K^+ concentration (8.8–16.0 mmol/l) evoked epileptic activity in the hippocampus (Zuckermann and Glaser 1968). Studies using K^+-ISMs showed that $[K^+]_e$ rose to a ceiling value of 8–12 mmol/l during a seizure (Hotson et al. 1973; Lux 1974; Moody et al. 1974; Sypert and Ward 1974; Lothman et al. 1975; Ookawa and Bureš 1975; Fisher et al. 1976a; Lothman and Somjen 1976; Ogata et al. 1976; Pedley et al. 1976b). If this ceiling level is exceeded, seizure activity is replaced by spreading depression (see Sect. 4.3.6.3).

Although seizure activity is always accompanied by an increase in $[K^+]_e$, this cannot be the only mechanism either of the development of epileptic activity, or of the transition between individual seizures and the end of a seizure. The increase in $[K^+]_e$ comes after the appearance of seizure activity rather than before it and, so far, no particular level of $[K^+]_e$ has been proven to have a causal relationship with the development of epileptic activity. In some cases, paroxysmal activity actually started when the $[K^+]_e$ was lower than the normal resting K^+ level (Lux 1974; Prince 1978). Impairment of the glial spatial buffer mechanism may also play a role in the development of seizures. Pollen and Trachtenberg (1970) suggested that deficiency of the glial buffer mechanism could result in an increase in $[K^+]_e$ and seizure activity. Some authors have reported no change in K^+ clearance in gliotic cortex (Glötzner 1973; Pedley et al. 1976a), whereas others have found slower clearance (Lewis et al. 1977).

In addition to paroxysmal neuronal activity, a depolarization shift of 20–30 mV is a characteristic feature of acute epileptiform foci (Prince 1978). These depolarization shifts represent either "giant" EPSPs, or they are caused by intrinsic regenerative neuronal events (Prince 1978). On the other hand, an increase in $[K^+]_e$ to 8–12 mmol/l also affects neuronal functions and is sufficient to produce a depolarization shift and epileptiform potentials (Somjen 1975; Prince 1978). Another important non-synaptic factor in the generation of seizures is a decrease in $[Ca^{2+}]_e$ to 0.9–0.5 mmol/l, which may also significantly affect neuronal excitability (Heinemann et al. 1977). Changes in the magnitudes of $[K^+]_e$ or $[Ca^{2+}]_e$ recorded in vivo evoked a statistically significant increase in excitability of neurones in hippocampal slice preparations (Schwartzkroin and Prince 1978; Prince and Schwartzkroin 1978; Prince 1978). In slices of the same tissue, an increase in $[K^+]_e$ to between 10 and 12 mmol/l caused epileptiform activity (Ogata 1975; Ogata et al. 1976; Prince et al. 1981). As soon as $[K^+]_e$ was raised to 5 mmol/l, the cells of these slices generated prominent depolarization shifts.

The mechanism by which an increase in $[K^+]_e$ can evoke epileptic seizures is still obscure. Wong et al. (1979) suggested that an increase in $[K^+]_e$ brought the neurones close to the threshold of an action potential burst by depolarizing the dendrites or modifying synaptic processes. Whistler and Johnson (1978) developed a computer model of epileptogenesis, involving slow membrane processes and $[K^+]_e$. The results obtained using this model point to a possible new role for $[K^+]_e$ in epileptogenesis, i.e., an increase in $[K^+]_e$ can transform stable cells to unstable (oscillating, endogenously active) cells, independent of synaptic activity. The massive release of transmitters, in association with high metabolic activity and ionic changes (K^+, Ca^{2+}, Na^+, Cl^-, and the pH changes) also leads to an increase in cell volume – which is due to the osmotic gradient directed into the cells and the diminution of the ECS – and all its consequences (see Chap. 7).

It can be concluded that although the increase in $[K^+]_e$ is a concomitant, rather than a causative, feature of seizure activity, the mechanism by which $[K^+]_e$ can influence epileptogenesis requires further study. A $[K^+]_e$ increment which outlasts the seizure's activity can significantly depolarize and even excite the terminals and synaptic regions of afferent fibres, or lead to their hyperpolarization and to an increase in evoked transmitter release. Finally, an increase in $[K^+]_e$ can also cause volume changes in the intra- and extracellular space. The increase in $[K^+]_e$ therefore represents an important feature involved in recurrent excitation, in hyperactivity, and in the time course of seizures.

4.3.6.3 $[K^+]_e$ and Spreading Depression

Leão's spreading depression (SD) (Leão 1944) is characterized by a complete blockade of neuronal activity which lasts several minutes, spreads through the tissue at a rate of about 3 mm/min, and is accompanied by a large negative extracellular potential of 20 mV or over (Bureš et al. 1974). This phase is usually preceded by a phase of hyperactivity. SD can be elicited by applying K^+, in a high concentration, to the surface of nervous tissue, or by injecting it into the relevant area. It can also be evoked by mechanical or chemical stimuli – such as the local administration of glutamate or ouabain, or dropping a metal object onto the surface of the exposed cerebral cortex. SD has been described in the cortex and in various other parts of the brain (but not in the spinal cord) of animals and man (Bureš et al. 1974; Šramka et al. 1977/1978).

There are three main hypotheses describing the mechanism of SD development, none of which excludes the others. Some authors postulate that SD is caused by the massive, simultaneous release of inhibitory and excitatory transmitters (Somjen 1973; Mori et al. 1976; Tuckwell and Miura 1978; Nicholson et al. 1981). This mechanism would allow all mobile ions to travel through subsynaptic channels along their electrochemical gradients. Van Harreveld (1959) suggested that the release of endogenous glutamate and K^+ from the cells changes the permeability of neuronal membranes, with resultant entry of Na^+ and Cl^-, accompanied by water which would follow them to establish osmotic balance; this causes the cells to swell, reduces the size of the ECS and increases brain impedance (see Chap. 7). Grafstein (1956) was the first to draw attention to the possible role of K^+ accumulation in the ECS as the outcome of a tremendous increase in neuronal activity resulting from massive stimulation of various brain structures in SD. The accumulated K^+, in turn, depolarizes neuronal elements and thus triggers further K^+ release. Radiotracer studies (Brinley et al. 1960b; Křivánek and Bureš 1960) revealed an enormous K^+ efflux from the cerebral cortex during SD, on almost the same scale as that seen in complete anoxia. Measurement with K^+ ISMs confirmed that SD

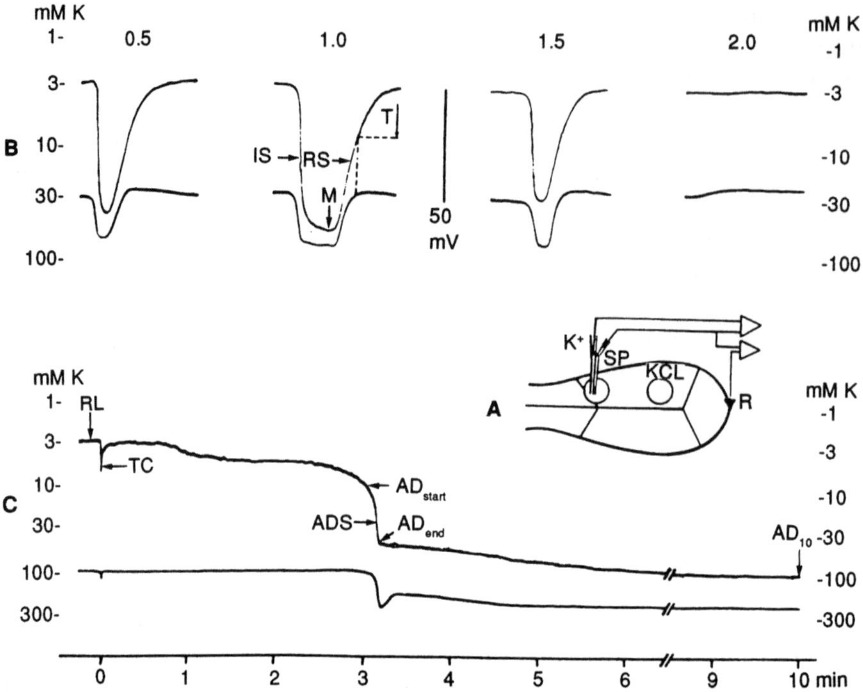

Fig. 44A–C. Measurements of extracellular brain potassium during SD and anoxic depolarization in the rat. **A** Experimental arrangement. *K*$^+$ and *SP*, K$^+$- and NaCl-channel of ISM, respectively; *R*, calomel reference electrode. **B** SD at different depths of the cortex (0.2–2.0 mm). Upper oscillograms correspond to K$^+$ electrode potential, lower records to the steady potential recording. *IS*, initial slope of the [K$^+$]$_e$ increase; *M*, maximum [K$^+$]$_e$; *RS*, slope of recovery during [K$^+$]$_e$ decrease; *T*, [K$^+$]$_e$ at the moment when the steady potential reaches the initial level. **C** Recording of the [K$^+$]$_e$ changes (*upper trace*) during anoxic depolarization (*lower trace*). *RL*, initial resting level of the cortical [K$^+$]$_e$; AD_{start}, onset of the rapid [K$^+$]$_e$ increase; *ADS*, slope at the steepest part of the [K$^+$]$_e$ increase; AD_{end}, transition of the rapid [K$^+$]$_e$ increase into the slow final phase; AD_{10}, [K$^+$]$_e$ 10 min after administration of *d*-tubocurarine (*TC*). (From Vyskočil et al. 1972)

usually begins with neuronal hyperactivity and an increase in [K$^+$]$_e$ to between 10 and 12 mmol/l (Fig. 44). In the second phase [K$^+$]$_e$ continues to rise until it has reached a level of between 30 and 80 mmol/l (Vyskočil et al. 1972; Prince et al. 1973; Futamachi et al. 1974; Mayevsky et al. 1974, Sugaya et al. 1975). This increase is accompanied by a negative shift in extracellular potential, cessation of all electrical activity, and massive depolarization of neurones and glial cells.

The increase in [K$^+$]$_e$ is also accompanied by changes in the extracellular Cl^-, Na^+, and Ca^{2+} concentration (Fig. 45). A dramatic drop of 50–100 mmol/l in Cl^- and of a 100 mmol/l in [Na$^+$]$_e$ were found in rat cerebral cortex (Hansen and Zeuthen 1981). These dramatic ionic changes are undoubtedly accompanied by water movements following the direction of osmotic gradients and by

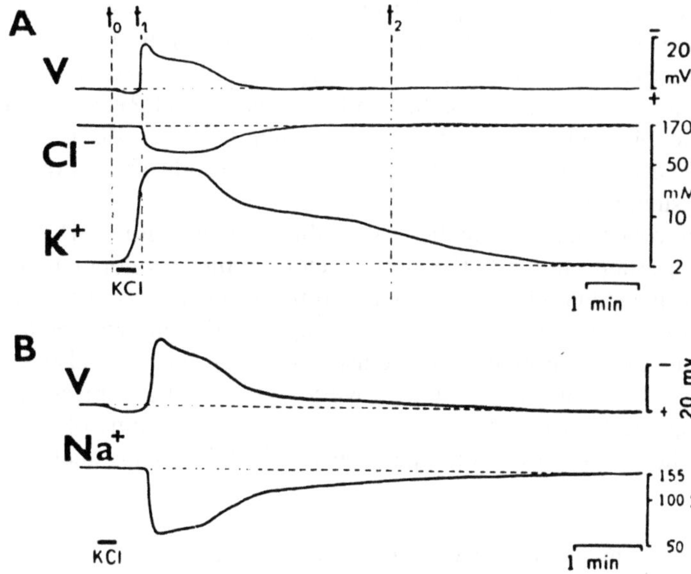

Fig. 45A,B. Cl^-, K^+, and Na^+ activity changes measured during SD in catfish cerebellum. **A** Simultaneously recorded slow potentials, ion signals, and field potentials. *Upper trace* (*V*) shows slow potentials during SD elicited by KCl microinjection (*KCl*). *Middle trace* shows a $[Cl^-]_e$ decrease and *lower trace* an increase in $[K^+]_e$. Between t_0 and t_2 a KCl microinjection was given. Between t_1 and the t_2 the characteristic SD occurs. (From Nicholson et al. 1977). **B** Simultaneous recording of slow potential (*V*) and $[Na^+]_e$ decrease during SD elicited by KCl microinjection (*KCl*). (From Kraig and Nicholson 1976)

swelling of the cells, as suggested by Van Harreveld (1959). The size of the ECS is halved (Phillips and Nicholson 1979; Nicholson et al. 1981). In addition, the $[Ca^{2+}]_e$ falls to between 10% and 50% of the resting value, as demonstrated in the cerebral cortex and cerebellum (Nicholson et al. 1977, 1978; Hansen and Zeuthen 1981). These ionic changes indicate that during SD neurones and neuroglia lose their capacity to maintain a transmembrane ion gradient. In this respect SD bears a striking resemblance to ischaemia.

According to the K^+ hypothesis of SD, SD propagation is mediated by diffussion of K^+ from cortex which is already depressed. When the increasing $[K^+]_e$ depolarizes neurones to the discharge threshold, more K^+ is released and the SD wavefront advances. Koroleva and Bureš (1979, 1980, 1982) showed that SD does not penetrate into electrically stimulated cortical regions in rats, which may result in SD reverberation around the focus. It has been found that repetitive cortical stimulation is accompanied by an increase in cortical $[K^+]_e$ to between 8 and 10 mmol/l, but a further rise above this ceiling level is prevented by enhanced active transport of K^+ into the neurones and by its uptake by glial cells (see Sect. 3.3). A similar increase in cortical $[K^+]_e$ after stimulation induced a SD blockade (Reddy and Bureš 1980). These findings led to the assumption that faster K^+ clearance disrupts the autoregenerative $[K^+]_e$

increase and thus stops SD propagation; the K^+ ions mediating the propagation of the SD wavefront are removed so rapidly that they do not trigger the positive feedback mechanism of SD. The increased activity of Na/K ATPase found in the area of the blockade (Křivánek and Reddy 1980) and the fact that the blockade persisted for a few minutes after $[K^+]_e$ had returned to, or even dropped below, the resting level (Reddy and Bureš 1980) support this assumption.

In conclusion, it can be stated that the increase in $[K^+]_e$ during SD is an important, but not the only, factor in the development of SD. Membrane permeability changes, massive transmitter release triggered directly or indirectly by ionic disturbances, changes in the energy metabolism of the brain, and disturbances of ionic redistribution mechanisms also accompany SD and affect its time course and recovery. Which of these mechanisms is the triggering mechanism which leads to SD in some regions of the brain, but not in others is, however, one question which still remains unanswered.

4.4 Dynamic K^+ Changes in the Organ of Corti

A number of findings on the K^+ accumulation in the brain and spinal cord have also been confirmed in the receptor organs. The findings and their significance do, however, warrant separate discussion in this and the following chapter since there are differences both in the mechanism of the K^+ changes and their functional significance.

4.4.1 Resting K^+ Concentration in the Inner Ear

Receptor auditory cells have a unique microenvironment. The apical surface of the hair cells is immersed in a fluid with a high K^+ content – the endolymph in the scala media. Endolymph contains about 140 mmol/l K^+ (Smith et al. 1954; Johnstone et al. 1963; Melichar and Syka 1977). This high K^+ concentration is maintained by means of an electrogenic pump in the cells of the stria vascularis (Sellick and Bock 1974; Melichar and Syka 1987). The basolateral surface of the hair cells is, however, surrounded by the fluid in the organ of Corti, i.e., by the cortilymph, which has a similar K^+ concentration to the perilymph (Syková et al. 1987b; Johnstone et al. 1989). Figure 46 shows the measurement of $[K^+]_e$ in the organ of Corti. A double-barrelled K^+-ISM was introduced into the cochlea through an opening in the bulla tympanica. In the perilymph of the scala tympani K^+ concentration is about 3 mmol/l (Melichar and Syka 1977; Johnstone et al. 1989). Johnston et al. (1989) found that the penetration of the ISM across the basilar membrane into Corti's tunnel (into cortilymph) was always accompanied by a $[K^+]_e$ increment of 0.4 ± 0.2 mmol/l ($n = 8$). During penetration of the microelectrode through the organ of Corti, whether towards the inner or the outer hair cells, no differences were found in the resting $[K^+]_e$. A rapid transient increase in $[K^+]_e$ occurs only during the impalement of

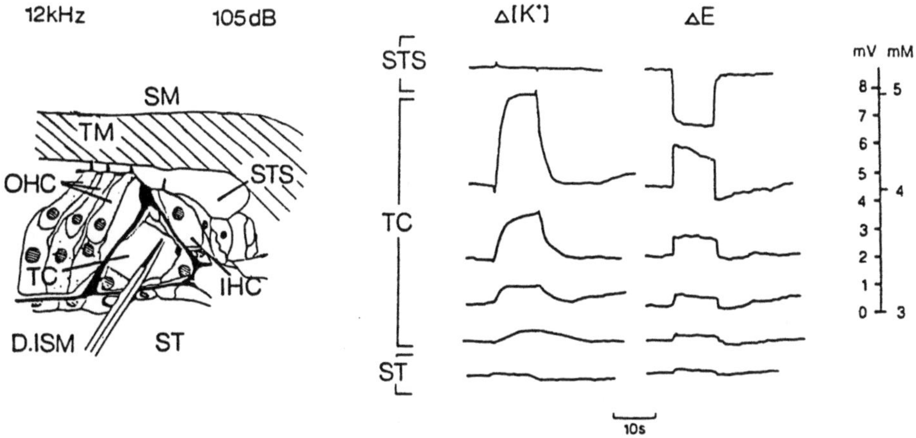

Fig. 46. Depth profile of changes in $[K^+]_e$ ($\triangle[K^+]$) and the extracellular dc receptor potential ($\triangle E$) in the guinea-pig organ of Corti during 10 s stimulation with 12 kHz at 105 dB. *Left*, organ of Corti, showing the direction of introduction of the K^+-ISM (*D. ISM*). The microelectrode penetrated the scala tympani (*ST*), crossed the basilar membrane into the tunnel of Corti (*TC*) and passed through the supporting cells and the inner hair cells (*IHC*) into the subtectorial space (*STS*). *OHC*, outer hair cells; *SM*, scala media. (From Johnstone et al. 1989)

hair cells or supporting cells. This increase is accompanied by a negative dc potential, which can be recorded with the reference channel of the microelectrode and thereby signalizes when a hair cell or a neighbouring supporting cell is impaled. When the microelectrode is pushed further through the cuticular plate, the dc potential rises to a positive endocochlear potential of about +80 mV with a simultaneous increase in $[K^+]_e$ to about 140 mmol/l. If the microelectrode is pushed deeper into the scala media, no further changes occur in either endocochlear potential or $[K^+]_e$.

4.4.2 Dynamic Changes in K⁺ Concentration in the Organ of Corti Evoked by Acoustic Stimuli

Stimulation with pure tone acoustic stimuli over a frequency range of 500 Hz–25 kHz produced changes in the potential and K^+ level in the ECS in the organ of Corti (Johnstone et al. 1989). These sound-evoked changes varied according to the location of the electrode tip within the organ of Corti and the intensity, frequency, and duration of the acoustic stimulus. The smallest $[K^+]$ changes were observed when the electrode was furthest away from the hair cells, i.e., when it was just below the basilar membrane. The increase in $[K^+]$ during stimulation is greater when the microelectrode is directed towards the hair cells. The maximum $[K^+]$ increment was seen in the immediate vicinity of the inner hair cells when the microelectrode passed through one supporting cell into the intercellular cleft between this cell and an inner hair cell (Fig. 46). A

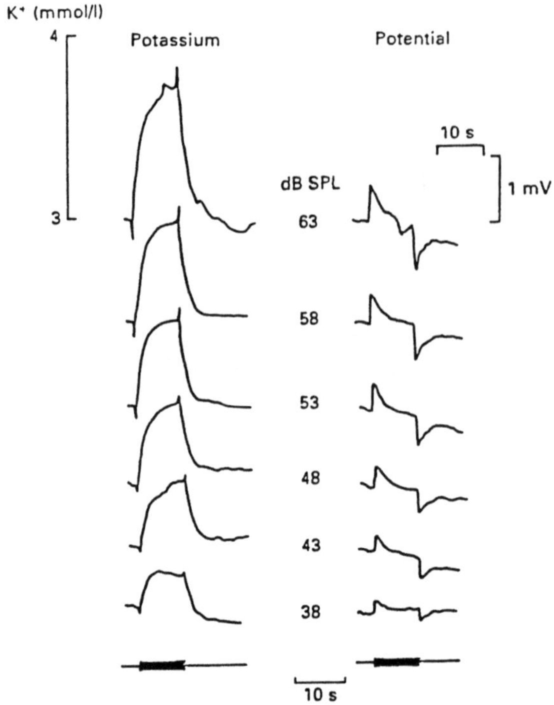

Fig. 47. Sound-evoked changes in K^+ level and extracellular potential in response to an 18 kHz tone burst presented at intensities between 38 and 63 dB SPL. Occurrence of the 18 kHz acoustic stimulus is indicated by a *thick bar* under the K^+ and potential waveforms. In this experiment the tip of the electrode was close to an inner hair cell. The abrupt transients at the start and end of stimulation are artifactual, and result from the slightly different response times of the reference and K^+ channels of the K^+-ISM. (From Johnstone et al. 1989)

similar depth profile was found when the microelectrode was introduced into the neighbourhood of outer hair cells, but, even in close vicinity to them, the $[K^+]$ changes were much smaller than those in the vicinity of inner hair cells – probably due to the fact that outer hair cells lie freely in cortilymph and are not surrounded by a layer of supporting cells. No changes in $[K^+]$ were observed when the microelectrode was situated in the scala media.

During stimulation with and relatively brief stimuli (10 s) at frequencies above 4 kHz, and with sound intensities lower than 80 dB SPL, the K^+ level in the tunnel of Corti and immediately adjacent to the inner hair cells increased steadily during the first 3 s and then remained relatively stable for the duration of the stimulus (Johnstone et al. 1989). After cessation of the stimulus the K^+ level quickly decreased to the pre-stimulation level with a time constant between 2 and 3 s (Fig. 47). In some cases, however, the K^+ was not completely cleared after acoustic stimulation and the level remained elevated (Fig. 48). The time course of sound-evoked changes in the extracellular potential

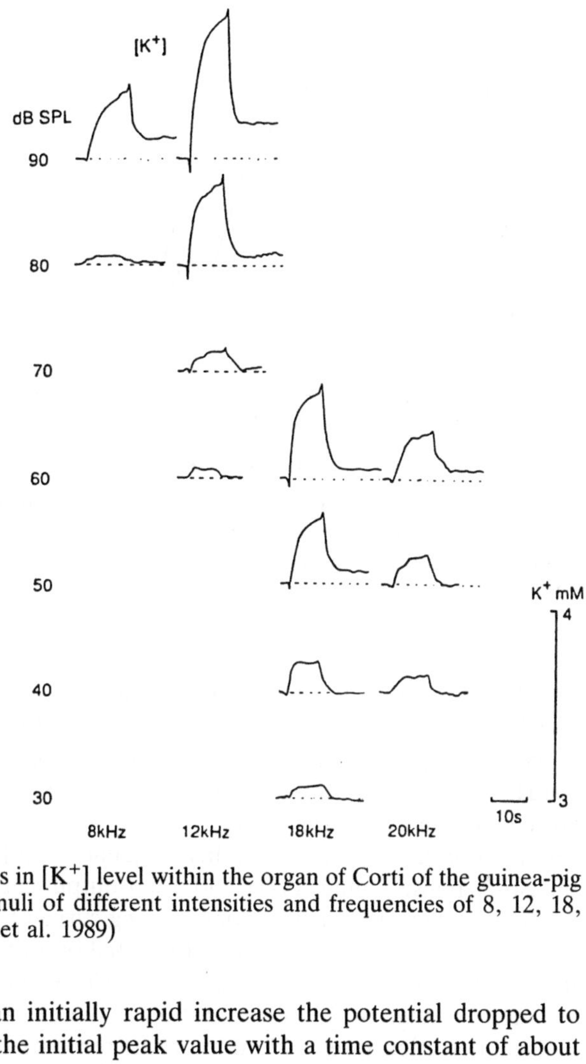

Fig. 48. Sound-evoked changes in [K^+] level within the organ of Corti of the guinea-pig as produced by pure tone stimuli of different intensities and frequencies of 8, 12, 18, and 20 kHz. (From Johnstone et al. 1989)

was quite different: after an initially rapid increase the potential dropped to approximately one-fifth of the initial peak value with a time constant of about 2–3 s. On cessation of the stimulus the potential fell transiently below its initial resting level, but then increased to the pre-stimulus level with a time constant of about 2 s (Fig. 47).

Stimulation with frequencies above 4 kHz raises [K^+] in the cortilymph, whereas stimulation with a frequency of 2 kHz or less causes it to fall below the resting value, accompanied by a negative extracellular dc receptor potential. In some cases, [K^+] changes were biphasic in character, i.e., [K^+] first fell and then rose. The results indicate that the [K^+] changes have two sources, one leading to an increase and the other to a decrease in K^+. The decrease observed during low frequency stimulation probably results from the reduction of K^+ currents flowing through the outer hair cells (Johnstone et al. 1989).

The dependence of K^+ and potential responses on stimulus intensity and frequency can be seen in Fig. 48 where stimulus frequencies of 8, 12, 18, and

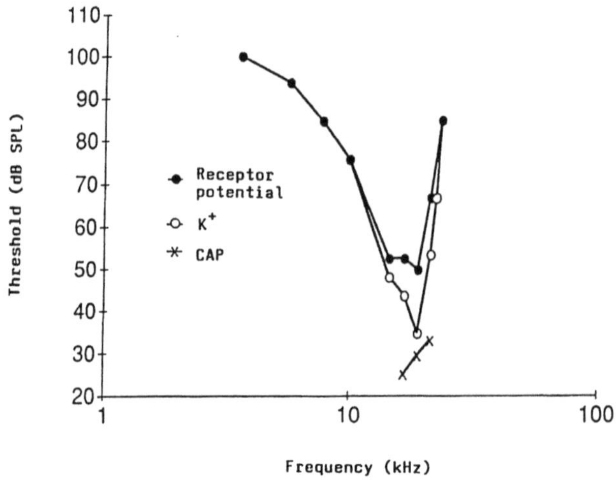

Fig. 49. Potassium (*open circles*) and extracellular receptor potential (*black circles*) iso-response tuning curves for stimulation frequencies in the critical frequency range. Compound auditory nerve action potential (*CAP*) threshold is also shown by the *dashed line*. (From Johnstone et al. 1989)

20 kHz were used with an intensity range of 30–90 dB SPL. It can be observed that K^+ and potential responses were elicited at lower sound intensities over a range of frequencies depending, of course, on the electrode location (18–20 kHz region) according to the neural tonotopic map for the guinea-pig (Robertson and Johnstone 1979).

The relationship between stimulus frequency and $[K^+]_e$ level is represerved in the iso-response tuning curves and input-output curves shown in Fig. 49. Figure 49 shows a typical K^+ iso-response tuning curve of the sound intensity required to produce a 0.05 mV change in differential output of the K^+-ISM when placed immediately adjacent to inner hair cells. This voltage change corresponded to a 10 μmol/1 change in K^+ concentration in the extracellular fluid, and presented a visual detection threshold on the chart recorder trace. K^+ level tuning curves were measured in many animals and they were all similar to the one shown in Fig. 49. Characteristic frequencies were between 16 and 21 kHz; low-frequency slopes ranged from 40 to 60 dB/octave ($\bar{x} = 50 \pm 14$ dB/octave) while high-frequency slopes ranged from 130 to 260 dB/octave ($\bar{x} = 205.1 \pm 45$ dB/octave).

During the first loud acoustic stimulus, the K^+ level increases by about 1 mmol/l and reaches a maximum after about 5 s. The K^+ level then starts to decrease, and after cessation of stimulation it declines to a level about 0.2 mmol/l less than the pre-stimulation resting level. This undershoot persists for tens of minutes following traumatic exposure. The amplitude of the sound-evoked extracellular potential is also reduced following overstimulation. At the outset of overstimulation the extracellular potential rises rapidly, but it then falls with a time constant of approximately 2–3 s during stimulation to a level

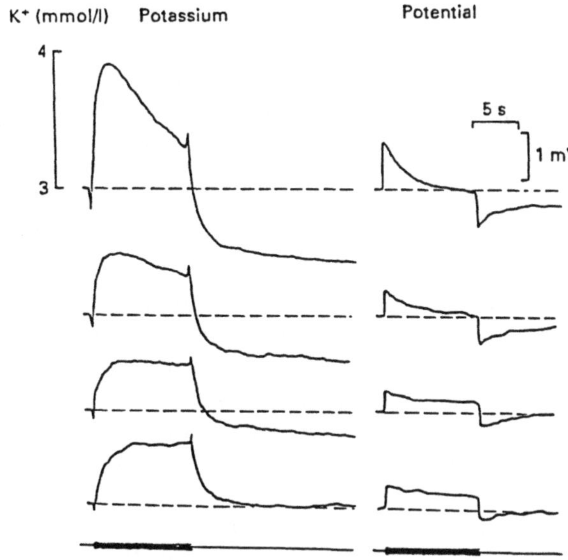

Fig. 50. Changes in K^+ level and extracellular potential produced by repeated stimulation with a 12 kHz tone burst lasting 30 s presented at 105 dB SPL. Occurrence of the acoustic stimulus is indicated with a *thick bar* beneath the curves. From *top*: 1st, 2nd, 4th, and 6th stimulations, respectively. Recordings in the organ of Corti of guinea-pig. (From Johnstone et al. 1989)

close to its pre-stimulation level. On cessation of the stimulus the extracellular potential transiently falls below the pre-stimulation resting level, returning only very slowly to that level.

Repetitive loud acoustic stimulation produces smaller peak K^+ and potential responses than those obtained for the first intense stimulus, and the adaptation of the potential and K^+ responses during the stimulus becomes less evident with each subsequent overstimulation (Johnstone et al. 1989). The initial exposure to acoustic overstimulation (12 kHz, 105 dB SPL for 30 s) resulted in elevation of the compound action potential threshold of about 15 dB at 18 kHz (i.e., at the frequency corresponding to the site of electrode insertion), and further exposures resulted in accumulated elevation of neural thresholds of 40–50 dB (Fig. 50).

$[K^+]$ also increases in the presence of moderate acoustic intensities (30–90 dB) but the time course and changes in $[K^+]$ and extracellular potential responses are different from those observed during acoustic overstimulation. In guinea-pigs, exposure to a strong acoustic stimulus which produces a temporary shift of auditory threshold probably leads to a reduction in resting K^+ currents flowing through the hair cells – thereby giving rise to undershoot below the resting $[K^+]$ level. There is evidence to show that K^+ plays a dominant role in the formation of permanent currents through hair cells (Crawford and Fettiplace 1981; Russell 1983).

Measurements of dynamic $[K^+]$ changes in the inner ear clearly demonstrate that acoustic stimulation raises $[K^+]_e$ both in Corti's tunnel and in the narrow intercellular clefts between the hair cells, i.e., in a space resembling the ECS of the CNS. Regarding the redistribution of K^+, it is possible that the supporting cells play a similar role to the glial cells in the CNS (see Sect. 4.4.3).

4.4.3 Functional Significance of Dynamic $[K^+]_e$ Changes in the Organ of Corti

There is already strong evidence to suggest that K^+ plays a dominant role in standing and stimulus-evoked currents through hair cells. The reversal potential of receptor potentials in the hair cells of the turtle cochlea (Crawford and Fettiplace 1981) and in guinea-pig inner hair cells (Russell 1983) is very similar to the endocochlear potential in these animals, in both of which K^+ is the major ion to carry the receptor current through the cells. Measurements of stimulus-related K^+ level changes in the organ of Corti support this view.

In the same way that K^+ accumulation in the CNS affects neuronal and glial cell function, it significantly influences the function of hair cells and supporting cells in the organ of Corti. For example, recent research shows that that there is a change in the shape of hair cells exposed to a raised $[K^+]_e$ (Brownell et al. 1985; Zenner et al. 1985).

The significance of reversible changes in the K^+ level in the organ of Corti is not clear. It is certainly expected that elevation of the $[K^+]_e$ depolarizes the cells of the organ of Corti. Cody and Russell (1985) have shown that the outer hair cell membrane potential depolarizes more slowly than that of adjacent inner hair cells during loud acoustic stimulation. Moreover, they demonstrated that when insufficient recovery time is allowed between repetitive acoustic stimuli, accumulated depolarization of the outer hair cells also occurs. Cody and Russell supposed that the decrease in the outer hair cell membrane potential resulted from the imbalance of ion content within the outer hair cell. Johnstone et al. (1989) suggest that accumulated depolarization during overstimulation, and outer hair cell depolarization immediately after loud acoustic stimulation may be the result of an increase in the K^+ level in the extracellular fluid surrounding the outer hair cells. The 2 mV depolarization observed by Cody and Russell in the outer hair cell resulted from an increase in $[K^+]_o$ of only 0.25 mmol/l. This value can easily be obtained with acoustic stimuli of a very low intensity. Of significance is the fact that changes in the membrane potential of outer hair cells during and after acoustic stimulation must contain a component which results from the elevation of extracellular K^+; these changes do not indicate impairment of the outer hair cells per se. The mechanism of the temporary threshold elevation which follows acoustic stimulation is not fully understood, although findings of Johnstone et al. (1989) help contribute to an understanding of the process. The role played by $[K^+]$ changes in reducing the sensitivity of the organ of Corti after acoustic trauma requires further research.

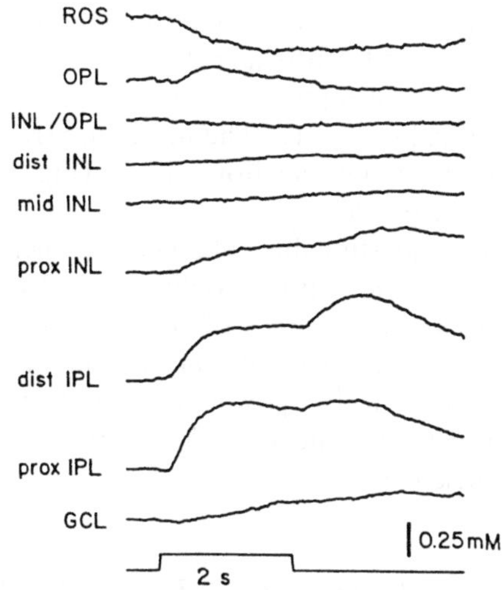

Fig. 51. Depth profile of light-evoked changes in K^+ concentration from retinal slice preparation. The baseline K^+ concentration was the same in the slice as in the superfusate – 2.5 mmol/l. *ROS*, rod outer segments; *OPL*, outer plexiform layer; *INL*, inner nucleus layer; *IPL*, inner plexiform layer; *GCL*, ganglion cell layer (From Karwoski et al. 1985)

4.5 Changes in K^+ Concentration in the Retina

Extracellular ionic changes in the retina of vertebrates (including mammals) and insects (the honeybee) have been studied by means of ISMs. $[K^+]_e$ changes produced by visual stimuli are of considerable interest, since they evidently contribute to the formation of the light-evoked potentials which make up the electroretinogram (ERG). The origin and clinical significance of the ERG was summarized by Armington (1974), and the origin of its individual waves was discussed by Tomita and Vanagida (1981). It has been found that when the $[K^+]_e$ is altered, the transmission of information is affected (Dowling and Ripps 1976).

Neuronal activity evoked by a visual stimulus produces a characteristic series of $[K^+]_e$ changes in the vertebrate retina. Oakley and Green (1976) showed a complex pattern of light-evoked K^+ flux in the retina. In the distal retina, light induces a decrease in $[K^+]_e$ with a time course corresponding to the c wave, while in the proximal retina and increase in $[K^+]_e$ can be seen. In the dark-adapted retina, hyperpolarization of the rods causes a prolonged decrease in $[K^+]_e$, owing to a K^+ shift to the inner segments (Oakley and Green 1976; Steinberg et al. 1980). Accumulation of K^+ in the inner and outer plexiform layer accompanies activity in depolarized bipolar, amacrine, and ganglion cells (Kline et al. 1978; Dick and Miller 1985; Karwoski et al. 1985). The increase in $[K^+]_e$ in the frog retina has a characteristic depth profile (Fig. 51; for a review see Karwoski and Proenza 1987). The largest and most thoroughly studied K^+

changes are those found in the subretinal space (among photoreceptors and pigmented epithelial cells), where the onset of photic stimulation reduces $[K^+]_e$ by several tenths of a mmol/l (Oakley and Green 1976); in other parts of the retina $[K^+]_e$ rises. The maximum $[K^+]_e$ increment (up to 0.5 mmol/l) was found in the inner plexiform layer (Fig. 51; Karwoski et al. 1985; Karwoski and Proenze 1987).

It is questionable whether the increase in $[K^+]_e$ has only one source, i.e., whether K^+ diffuses only proximally and distally in the retina, or whether there are several sources of $[K^+]_e$ at different depths. The latter eventuality seems more likely (see Karwoski and Proenza 1987). It is also important that an increase in $[K^+]_e$ was found not only at the onset of the visual stimulus, but also at the end (Fig. 51; Karwoski et al. 1985; Karwoski and Proenza 1987; Dick 1979).

4.5.1 Regulation of $[K^+]_e$ by Glial Cells in the Retina

Activity-evoked $[K^+]_e$ changes in the retina are carefully buffered, so as to limit their effect on retinal function to the minimum (for a review see Newman 1987). It was found that small changes in $[K^+]_e$ also produce small changes in the neuronal resting potential and that these lead to changes in the effectiveness of synaptic transmission and in the threshold for the formation of action potentials. As in the CNS, a number of mechanisms participate in the buffering of $[K^+]_e$ changes in the retina (see Chap. 3). The relative importance of mechanisms participating in K^+ homeostasis in the retina is not clear, but there is evidence that Müller's cells, which are actually the glial cells of the retina and are in many respects related to astrocytes, play an important role. Müller's cells are radial glial cells whose processes pass through the entire depth of the retina, from the inner to the outer limiting membrane. From the cell body, which lies in the inner nuclear layer of the retina, two radial processes (one proximal, one distal) project into the vitreous body to the level of the inner photoreceptor segment. Like other glial cells, Müller's cells are mainly permeable to K^+ (Fig. 52; Newman 1985a; Conner et al. 1985). Their K^+ selectivity and their elongated shape are ideal for their function as cells assuring the spatial buffer mechanism first described in the honeybee's retina (Coles and Orkand 1983).

The retina of the honeybee drone is composed of photoreceptors and glial cells. Coles and Tsacopoulos (1979) described K^+ dynamics in the drone retina by monitoring $[K^+]_e$ in the retina and $[K^+]$ in photoreceptors and glial cells (Fig. 53). They found that drone receptors were depolarized in response to light, due to an influx of Na^+ and an efflux of K^+. K^+ efflux from photoreceptors resulted in a substantial $[K^+]_i$ decrease, but only a small $[K^+]_e$ increase. In glial cells, however, $[K^+]_i$ rose significantly. Gardner-Medwin, Coles, and Tsacopoulos (1981) studied $[K^+]_e$ increases and field potentials in the drone retina and demonstrated that a light-evoked increase in $[K^+]_e$ established a current flow which entered glial cells in regions with a high $[K^+]_e$

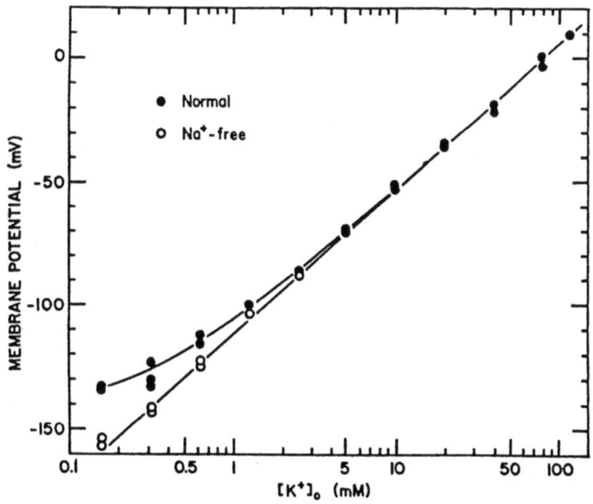

Fig. 52. Müller cell membrane potential as a function of perfusate $[K^+]$. All data is recorded from a single cell from a frog retinal slice. The cell is almost exclusively permeable to K^+. In normal perfusate (*filled circles*) E_m derivates from the K^+ equilibrium potential (*straight line*) only when $[K^+]$ is lower than 2.5 mmol/l. In Na-free perfusate (*open circles*), E_m follows E_k faithfully. (From Newman 1985c)

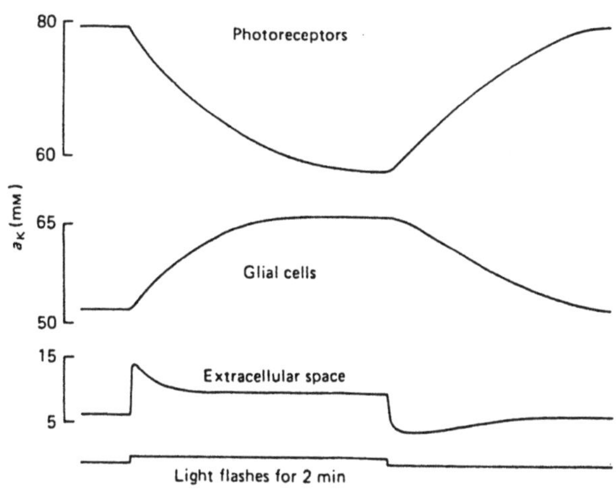

Fig. 53. Summary of K^+ level changes in the superfused, cut retina of the honeybee drone during light stimulation. Efflux of K^+ from the photoreceptors causes a decrease in intracellular $[K^+]$ in these cells (*top*). Almost all of the K^+ released into ECS (approximately 99%) enters the glial cells, causing a large increase in $[K^+]$ in the glial syncytium (*middle*). Both spatial buffering currents and net uptake mechanisms are responsible for the influx of K^+ into glial cells. Entry of K^+ into these cells limits the rise in $[K^+]$ in ECs (*bottom*) to a relatively small increase. K^+ levels are shown as K^+ activities (a_k) and represent the mean values measured at each recording site. These activities can be converted to concentrations by multiplying by 1.43. (From Coles and Tsacopoulos 1979)

and left the glial syncytium in regions where $[K^+]_e$ was lower (see Fig. 6). The results suggest that a significant portion of K^+ released from photoreceptors enters the glial syncytium and is attributable to spatial buffer currents.

In the vertebrate retina, the modern conception is that a visual stimulus causes hyperpolarization of photoreceptors and a decrease in $[K^+]_e$ in their vicinity. This decrease triggers the spatial buffer mechanism in Müller's cells, so that K^+ leaves the cells in areas with a low concentration and enters them in areas with a raised $[K^+]_e$ concentration, i.e., a current loop is formed through the ECS, from the sclera to the vitreous body, where it gives rise to a negative potential in the vitreous body (Zuckerman 1973; Hanitzsch 1973). Spatial buffer currents in the vertebrate retina and a current source-density analysis of the b wave of the ERG are shown in Fig. 7 (see Newman 1985a,b).

It can be concluded that $[K^+]_e$ changes in the retina probably do not act as a signal between receptor cells. A very stable microenvironment is very important to the receptor cells because they must respond to very small changes in the visual field. The glial cells in both vertebrate and invertebrate retinas are instrumental in clearing activity-related increases in $[K^+]_e$. Spatial buffer, and perhaps also uptake mechanisms, play an important role in this process and, as they do in the brain, participate in K^+ homeostasis.

5 Dynamic Changes in Extracellular Na^+, Cl^-, and Ca^{2+} Concentration

In addition to $[K^+]_e$ changes, neuronal activity is accompanied by changes in the concentration of other ions in the microenvironment of the CNS and receptor organs. In the case of Na^+ and Cl^- it is difficult to measure the changes directly, since the sensitivity of the relevant ISMs is relatively small in the high resting extracellular concentration of these ions. Changes in Na^+ and Cl^- can therefore only be measured when they exceed several mmol/l; they are thus hard to demonstrate during electrical stimulation and impossible to demonstrate during adequate stimulation. However, the dramatic changes (i.e., the decrease in $[Na^+]_e$ and in $[Cl^-]_e$) which occur during anoxia, epilepsy, and spreading depression can be measured dynamically (see Fig. 45; Nicholson 1980; Nicholson et al. 1981).

The decrease in $[Ca^{2+}]_e$ due to neuronal activity may be essential for the modulation of transmission in the CNS. The decrease results from the movement of Ca^{2+} into neurones and presynaptic endings. A decrease in $[Ca^{2+}]_e$ has been observed during excitation in different parts of the brain and spinal cord, during electrical stimulation of afferent input, and in pathological states.

5.1 Changes Induced in Resting $[Ca^{2+}]_e$ During Stimulation of Afferent Input

The resting $[Ca^{2+}]_e$ in the brain and spinal cord of amphibians and mammals varies from 1.2 to 1.6 mmol/l. Repetitive electrical stimulation causes a 0.1–0.5 mmol/l drop in the resting level, i.e., $[Ca^{2+}]_e$ falls below 1.0 mmol/l (Nicholson 1980; Chvátal et al. 1988). In the cerebral cortex, cerebellum, thalamus, and hippocampus, $[Ca^{2+}]_e$ falls during repetitive electrical stimulation until it reaches the minimum level, which is about 0.8–0.9 mmol/l (Lux 1974; Heinemann and Lux 1977; Somjen 1979; Morris 1981; Krnjević et al. 1982a,b; Chvátal et al. 1988). When stimulation is discontinued, recovery takes place quickly – as does recovery from an increase in $[K^+]_e$ – and there is an "overshoot" to concentrations higher than the original resting level. The depth profile of stimulation-induced $[Ca^{2+}]_e$ changes in the cortex, cerebellum, thalamus, and spinal cord is identical to the depth profile of $[K^+]_e$ changes (Heinemann et al. 1977; Nicholson et al. 1976).

Figure 54 illustrates the typical decrease in $[Ca^{2+}]_e$ in the dorsal horn of the isolated frog spinal cord (Chvátal et al. 1988). During stimulation, the maximum decrease (up to 0.5 mmol/l) was found in the deeper layers. The decrease in $[Ca^{2+}]_e$ is always proportional to the intensity, frequency, and duration of stimulation. If, however, electrical stimulation of the dorsal spinal root lasted longer than 20–30 s, there was no further change in $[Ca^{2+}]_e$, i.e., the ceiling

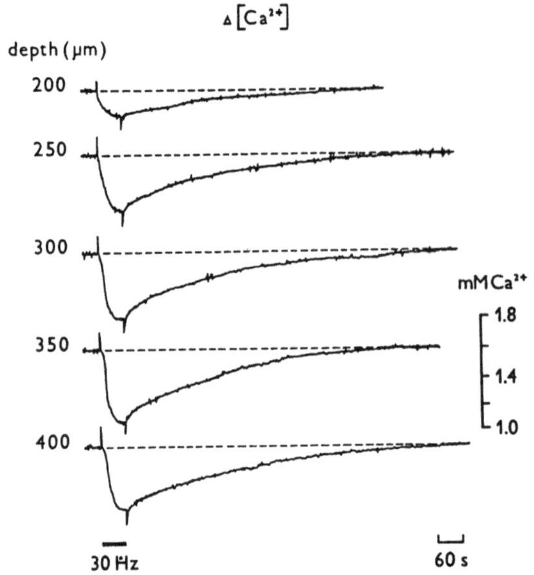

Fig. 54. Depth profile of $[Ca^{2+}]_e$ changes (*$\Delta[Ca^{2+}]$*) in the isolated frog spinal cord during repetitive electrical stimulation of dorsal root (30 Hz, 60 s). The maximum decrease in $[Ca^{2+}]_e$ was found in the deeper layers of the dorsal horns. (From Chvátal et al. 1988)

value had been reached. The time course of $[Ca^{2+}]_e$ recovery to the resting level was similar to that observed after $[K^+]_e$ and pH_e changes (see Fig. 61; Chvátal et al. 1988). As was the case with $[K^+]_e$ and pH_e, $[Ca^{2+}]_e$ changes in the ventral spinal horns were only slight.

5.2 $[Ca^{2+}]_e$ Changes in Pathological States

Whereas the $[K^+]_e$ has repeatedly been shown to increase during interictal and ictal discharges (see Sect. 4.3.6.2), the $[Ca^{2+}]_e$ decreases during such activity (Fig. 55; Heinemann et al. 1977, 1978, 1981; Prince 1978; Heinemann and Louvel 1983). An increase in dendritic permeability for Ca^{2+} may be significant in the development of epileptic activity; changes in epileptic activity are probably reflected in changes in $[Ca^{2+}]_e$. In cats, it was, in fact, demonstrated that the onset of paroxysmal activity after the administration of pentylenetetrazol, or in an epileptic focus, was preceded by a drop in $[Ca^{2+}]_e$. Seizure periods are always preceded by a decrease in $[Ca^{2+}]_e$ (Heinemann et al. 1977). A particularly pronounced decrease in $[Ca^{2+}]_e$ was found in the cortical layers adjoining a chronic epileptogenic scar. It is possible that dendritic permeability for Ca^{2+} is activated by the increase in $[K^+]_e$, since a decrease in $[Ca^{2+}]_e$ always follows close on the heels of such an increase. Pumain et al. (1983) found that individual epileptic action potentials were preceded by an abrupt

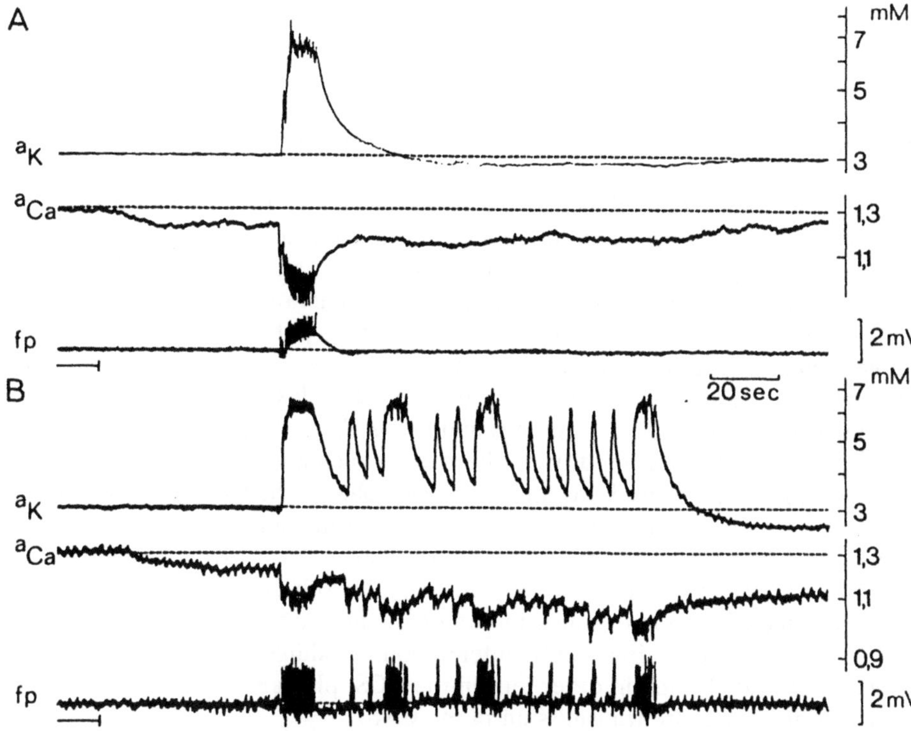

Fig. 55A,B. Simultaneously recorded changes in $[K^+]_e$ (a_k), $[Ca^{2+}]_e$ (a_{CA}), and focal potentials (*fp*) during and after injection of 10 mg/kg body-weight pentylenetetrazol. End of injection (duration 30 s to 3 min) indicated by *bars*. Recording depth of $[Ca^{2+}]_e$ is 300 μm in **A** and 1100 μm in **B**. Recording depth of $[K^+]_e$ in both **A** and **B** is 500 μm. (From Heinemann et al. 1977)

drop in $[Ca^{2+}]_e$. There may well be reciprocal activity, with the increase in $[K^+]_e$ activating dendritic permeability for Ca^{2+}; this in turn leads to the release of K^+, thereby contributing to the depolarization shifts which are a characteristic phenomenon of epilepsy.

Epileptiform activity was also induced in rat hippocampal slices by lowering $[Ca^{2+}]_e$ (Konnerth et al. 1986, 1987; Albrecht and Heinemann 1989; Albrecht et al. 1989). Perfusion of slices with a low Ca^{2+} concentration blocks the synaptic responses of hippocampal pyramidal cells. It has been concluded that the hippocampal CA1 area is capable of generating and sustaining maximal seizure activity in the absence of chemical synaptic transmission, but the mechanism of low Ca^{2+}-evoked seizure activity is not clear. This preparation does, however, serve as an in vitro model for studies of seizure activity.

Lastly, during ischaemia, severe hypoglycaemia, and anoxia in the mammalian cortex, $[Ca^{2+}]_e$ falls to very low values of about 0.1 mmol/l (Harris and Symon 1984; Harris et al. 1984). The changes are only partly reversible. A similar decrease was also observed in fish and rat cerebellum during spreading depression (see Fig. 45; Nicholson et al. 1977, 1978; Kraig and Nicholson 1978).

5.3 Functional Significance of Dynamic $[Ca^{2+}]_e$ Changes

In addition to the part played by $[Ca^{2+}]_e$ changes in the formation of slow potentials, epileptic seizures, spreading depression, and anoxia, stimulation-induced $[Ca^{2+}]_e$ changes can also display excitatory and inhibitory effects. The former include lowering of the threshold for the generation of action potentials and facilitation of the formation of ectopic action potentials. The inhibitory effects comprise a decrease in transmitter release. Regulatory mechanisms in the brain, however, have the task of preventing loss of extracellular Ca^{2+} in the vicinity of presynaptic endings and nerve fibres. If a marked decrease in $[Ca^{2+}]_e$ occurs, the excitability of the axones increases owing to a decrease in their threshold caused by local surface voltage changes (Frankenhauser and Hodgkin 1957). The efficacy of synaptic transmission decreases, however, when the electrochemical gradient for Ca^{2+} across the presynaptic membrane is lowered. Malenka et al. (1989) recently hypothesized that postsynaptic $[Ca^{2+}]_e$ affects synaptic transmission and that the sequence of events initiated by Ca^{2+} may lead to long-term potentiation. Findings to date indicate that changes in $[Ca^{2+}]_e$, and in other ions in the ECS, play an important role in the modulation of impulse transmission in the CNS.

6 Dynamic pH_e Changes

The pH value in the extracellular fluid is about 7.2–7.3, which means that the extracellular hydrogen ion concentration ($[H^+]_e$) is about 0.05 μM. The biological importance of stability of pH_i has long been known, but recent findings demonstrate that dynamic pH_e changes during neuronal activity may also influence the function of nerve and glial cells.

Homeostasis of pH_e is closely associated with changes in pH_i. The intra- and the extracellular spaces represent two separate compartements of the brain, however, each with its own physicochemical buffers, ion shifts, and metabolic processes which form acids and bases, i.e., agents which influence their pH.

Stewart (1978, 1981) submitted a conceptional plan for general resolution of the problem of acid-base relationships in biological fluids. The definition of the pH as the negative logarithm of the hydrogen ion concentration ignores the fact that H^+ is present in the brain in submolar concentrations. The definition also fails to make it clear that $[H^+]$ is a function of equilibrium of the dissociation of water, weak acids and bases, CO_2, carbonic acid, HCO_3^-, the solubility of carbonates, the law of electroneutrality, and the law of the preservation of mass. In order to determine $[H^+]$ in any compartment of the brain, we must take all these factors into account. Stewart differentiates between "independent" and "dependent" variables of the acid-base system. There are three independent variables:

1 [SID]: the strong ion concentration difference, i.e., the difference between the concentration of strong (base) cations and strong (acid) anions (this system includes ions which undergo complete dissociation, such as Na^+, K^+, Ca^{2+}, Mg^{2+}, Cl^-, and lactate).
2 $[A_{tot}]$: the total concentration of weak acids and bases, represented by various ionizable groups of proteins.
3 pCO_2: partial carbon dioxide pressure.

It is important to bear in mind that these three independent variables fully define the pH_e and other dependent variables in the ECS, such as $[H^+]$, $[OH^-]$, and $[CO_3^{2-}]$. In solutions containing no buffers, the pH is determined very strictly by [SID], since it is the only variable which can change the ionization of the water. If weak acids are present, the pH will also be affected by the amount of these acids. The pH_e is determined entirely by [SID] and the pCO_2 (owing to the formation of HCO_3^- and carbonic acid), since virtually no weak acids occur in the ECS. $[H^+]_e$ can thus be determined using the following equation:

$$[H^+]_e = \frac{K_i' \cdot S' \cdot pCO_2}{[SID]}$$

in which K_i' is the ionization constant of carbonic acid and S′ is the solubility constant of CO_2 (Kraig et al. 1985a). This equation is analogous to Henderson's classic equation (Henderson 1908), since [SID] is more or less equivalent to the extracellular [HCO_3^-] (Stewart 1981).

Stewart's approach, however, only describes how equilibrium states differ. Systemic – physiological pH shifts can be classified according to the altered independent variable. Systemic acidosis can develop as a result of an increase in plasma pCO_2 (respiratory acidosis) or a drop in the plasma's [SID] (metabolic acidosis). Activity-related pH_i and pH_e shifts accompany a transmembrane change in [SID]. However, the numerous membrane transport mechanisms result in changes in intracellular and extracellular [SID] and therefore in pH changes.

The original hypothesis which postulated that the pH_i of animal cells depends solely on the external pH and the membrane potential, and that H^+, OH^-, and HCO_3^- are distributed passively across cell membranes has been proved incorrect by research over the past few years (for a review see Thomas 1988b). We now know that there must be an intracellularly localized process which is responsible for the continuous clearance of acids, which keeps the pH_e more acid than the pH of the blood or CSF (see Fig. 57). The metabolic activity of cells in the nervous system is accompanied by intracellular acid shifts (a decrease in pH_i) which are due either to direct production of acid or CO_2. The mechanism of H^+ production may vary according to the type of cell – neurones, glial cells, or endothelial cells – but in the last instance excess H^+ is expelled from all cell types by one of the mechanisms already described or as yet unknown. Experiments on diverse types of cells have shown that the pH_i is kept alkaline by means of specific transport processes in the cell membrane (see Roos and Boron 1981; Nuccitelli and Deamer 1982; Thomas 1984a, 1988b). The transport mechanisms which participate in the pH homeostasis of animal cells are shown in Fig. 56.

In a number of tissues, including the vertebrate CNS, pH_i is regulated by an electroneutral Na^+/H^+ exchange system (Fig. 56a; Aickin and Thomas 1977; Boron and Boulpaep 1983; Deitmer and Ellis 1980; Grinstein et al. 1984; Moolenaar et al. 1983; Chesler and Nicholson 1985; Deitmer and Schlue 1987; Kaila and Vaughan-Jones 1987). This Na^+/H^+ exchange, which is blocked specifically by amiloride, appears to be the main mechanism for the removal of excess H^+ from cells, the source of energy for its removal being the transmembrane Na^+ gradient. It is important, for instance, that lowering of the Na^+ gradient after inhibition of the Na^+/K^+ pump leads to a drop in pH_i (Deitmer and Ellis 1980; Vaughan-Jones et al. 1983).

A mechanism which probably also utilizes the energy of the Na^+ gradient across the cell membrane to expel H^+ and Cl^- in exchange for Na^+ and HCO_3^- was demonstrated, in detail, in neurones of the snail's nervous system (Thomas 1977, 1984b; Evans and Thomas 1984). Thomas asserts that the introduction of a single Na^+ ion into a cell provides sufficient energy for the extrusion of two H^+ ions (or the extrusion of one H^+ and the uptake of one HCO_3^- ion). The release of H^+ (which is virtually inseparable from OH^-

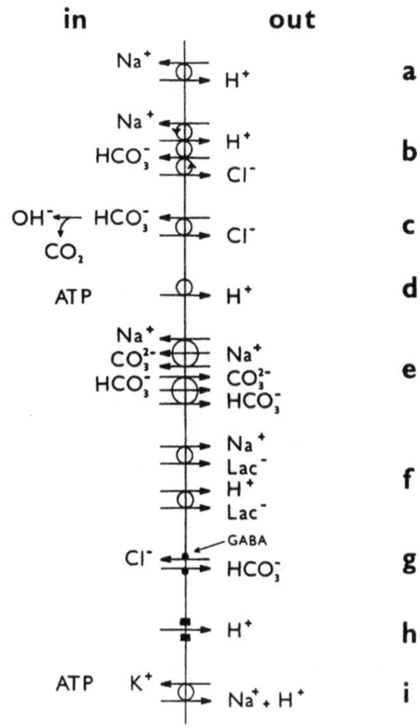

Fig. 56. Suggested transport mechanisms participating in pH homeostasis in animal cells. *a*, Na^+/H^+ exchange; *b*, Na^+/H^+-HCO_3^-/Cl^- cotransport; *c*, HCO_3^-/Cl^- exchange; *d*, the proton pump; *e*, Na^+-HCO_3^--CO^{2-} cotransport; *f*, lactate transfer; *g*, the Cl^- and HCO_3^- shift via the chloride channel; *h*, the proton channel; *i*, the Na^+/K^+ pump. (After Thomas 1988b)

uptake) is accompanied by an increase in intracellular Na^+ activity and a decrease in Cl^- activity, i.e., the "acid extrusion system". This exchange system is inhibited by the removal of external Na^+ or HCO_3^- and by specific inhibitors of Cl^-/HCO_3^- exchange: SITS (4-acetamide-4′-isothiocyano-2,2′-disulphostilbene), and DIDS (4,4′-diazidistilbene-2,2′-disulphonic acid). This pH_i regulatory system, which also appears to be electroneutral, is shown in Fig. 56b. A similar mechanism has also been described in squid axones and mollusc muscles (for a review see Roos and Boron 1981). It is interesting to note that this mechanism can be reversed, i.e., the ions can move in the opposite direction (the "acid uptake system"), whereby H^+ is carried into the cells with Cl^-, while Na^+ and HCO_3^- are transported out (Russell et al. 1983; Evans and Thomas 1984). This reversed system can also be blocked in a chloride-free medium by using SITS or lowering the pH_e. Reduction of the pH_e leads to elevation of intracellular Cl^- and a decrease in Na^+. These changes are the reverse of those seen when the pH_i regulatory system operates normally. Evans and Thomas (1984) suggested that extracellular acid changes might reverse the pH_i regulatory system by causing H^+ ions to be transported into the cells instead of being extruded from them.

Thomas (1976, 1984b) further stated that Na^+ which enters the cell in conjunction with H^+ extrusion is subsequently pumped out of the cell by an electrogenic pump (Fig. 56i) utilizing energy obtained from the hydrolysis of ATP. H^+ is thus expelled with Na^+, in exchange for K^+. The decrease in the ATP level does not necessarily block H^+ transport directly; it may be blocked as a result of a decrease in the amount of extracellular Na^+ (Moody 1981, 1984).

A further important mechanism that has been studied in detail in red blood cells and also in mammalian smooth muscle (Aickin and Thomas 1977) is simple Cl^-/HCO_3^- exchange, which can be blocked specifically by the stilbene derivates SITS and DITS (Fig. 56c). In the mouse soleus muscle this exchange amounts to only about 20%, however, and probably requires metabolic energy (Aickin and Thomas 1988). The same mechanism may also operate in astrocytes (Kimelberg et al. 1979a; Kimelberg 1981; Deitmer and Schlue 1987) and in other cells (see Roos and Boron 1981; Thomas 1984a).

Another suggested mechanism is the "hydrogen pump", or electrogenic H^+ pump, which utilizes ATP as the energy source (Fig. 56d). Proton-translocating ATPase is present in the kidneys (Gluck and Caldwell 1988), but its role in the CNS requires further research.

A further mechanism which has been described in various tissues and may also operate in nerve cells is Na^+-HCO_3^- cotransport in the ratio one part Na^+ to three parts HCO_3^-; the intermediate product in this mechanism is CO_3^{2-} (or $NaCO_3^-$), as illustrated in Fig. 56e (Boron and Boulpaep 1983; Frömter et al. 1988; Wang et al. 1987). This Na^+-dependent (and probably electrogenic) HCO_3^- transport into cells may also operate in glial cells (Deitmer and Schlue 1987; Schlue and Deitmer 1988; Astion and Orkand 1988; Kettenmann and Schlue 1988).

Metabolic acidosis associated with lactic acid synthesis is accompanied by the transport of lactate across cell membranes together with Na^+ and H^+ (Fig. 56f). This mechanism has been described in both the kidneys (see Boron et al. 1988) and in the CNS, in various pathological states (see Sect. 6.6).

Kaila and Voipio (1987) demonstrated that bicarbonates pass in the direction opposite through chloride channels activated by the inhibitory transmitter GABA during inhibitory synaptic processes (Fig. 56g). In this way GABA produces a dramatic drop in the pH_i owing to loss of HCO_3^- from postsynaptic elements and associated extracellular alkalosis. Bicarbonate ions which leave the cells via chloride channels are replaced intracellularly by a reaction in the presence of the excess CO_2 and H_2O, which is catalysed by carbonic anhydrase; this reaction gives rise to both HCO_3^- and H^+. One H^+ ion is formed for every HCO_3^- ion, so that the pH_i falls again. If Cl^- channels are re-opened by means of GABA, HCO_3^- leaves the cells via these channels and Cl^- enters them. Findings which indicate that Cl^- channels activated by the inhibitory transmitters GABA and glycine are also permeable for HCO_3^- (Bormann et al. 1987; Kaila and Voipio 1987) show that intracellular acidosis induced by GABA can be an important modulating mechanism in inhibitory transmission. These recent findings testify that all postsynaptic inhibitory ion shifts are accompanied by changes in the pH_i and pH_e.

Last, but not least, I should like to draw attention to the recent discovery of proton channels in snail neurones (Fig. 56h; Meech and Thomas 1987; Thomas 1988c). Thomas (1988a) described the accumulation of H^+ on the surface of the neuronal membrane of snail neurones during their depolarization by the voltage clamp method. Conversely, hyperpolarization can be accompanied by an increase in the pH_e owing to H^+ influx into the cell via this channel. The H^+ shifts along these proton channels may well play a role in pH_e changes observed in excitable tissues – including the narrow intercellular spaces in the vertebrate CNS.

In conclusion, I should like to emphasize that all the mechanisms so far suggested could participate in regulation of the pH_e. The question remains as to of whether the dynamic pH_e changes associated with neuronal activity result from changes in $[SID]_e$, or from specific transport systems, or whether H^+ is transported mainly by voltage-dependent H^+ channels.

6.1 Extracellular Buffering Power

Before describing dynamic changes in pH_e related to neuronal activity it is necessary to point out that acid-base shifts of considerable magnitude can be masked in well-buffered interstitial solution. Any physiological process which leads to pH_e changes is hindered by effective extracellular buffering. In the extracellular fluid, bicarbonate and CO_2 constitute the major buffer system since the concentration of proteins and organic acids is negligible (Fencl 1986). The HCO_3^- and CO_2 participate in buffering according to the reaction.

$$CO_2 + H_2O \overset{K_1}{\rightleftharpoons} H_2CO_3 \overset{K_2}{\rightleftharpoons} \underset{\substack{+\\ H^+}}{HCO_3^-} \overset{K_3}{\rightleftharpoons} CO^{2-}{}_3 + H^+$$

where K_1, K_2, and K_3 are the equilibrium constants for hydration and dissociation reactions. The reaction is speeded up by the presence of the enzyme carbonic anhydrase (CA) (see Sect. 6.3.5). At a physiological pH, the contribution of $CO^{2-}{}_3$ is negligible and the extracellular buffering power (β) can therefore be calculated approximately using the equation:

$$\beta = 2.3\,[HCO_3^-]$$

In mammals, assuming that the pH_e is about 7.3 and the pCO_2 46 torr, the calculated $[HCO_3^-]$ would be 22 mmol/l and the extracellular buffering power about 50 mmol/l. Since extracellular $[HCO_3^-]$ varies greatly in different animal species, the buffering power of the interstitial fluid varies accordingly. As the buffering power is about four times greater in terrestrial animals than in aquatic species, we can expect larger pH_e shifts in aquatic animals. Recent studies have shown that stimulation-evoked alkaline pH_e shifts in the rat cerebellum (Kraig et al. 1983) were about 5–10 times smaller than those in the skate cerebellum (Rice and Nicholson 1988).

6.2 Activity-Related Dynamic pH_e Changes in Nervous Tissue

Since the pH_e depends on depolarization of the neuronal membrane, on the metabolic activity of the nerve cells, and on ion shifts across the cell membrane, any neuronal activity is likely to be accompanied by dynamic changes in the pH. The measurement of dynamic pH_e changes in the brain and spinal cord in vivo was first made possible by the use of pH-ISMs with a liquid ion exchanger in their tip (Ammann et al. 1981). Authors had previously used glass pH-ISMs to show that dynamic changes occurred in the pH_e during stimulation (Urbanics et al. 1978) or SD (Lehmenkühler et al. 1981), but their results were not precise owing to the slow time constant, relatively large tip, and limited selectivity of the microelectrodes. The measurements were also not accurate because local nerve cell potentials could not be read simultaneously, as is now the normal procedure with double-barrelled pH-ISMs with a liquid ion exchanger in their tip (see Chap. 2).

6.2.1 Resting pH_e

The resting pH_e measured by various authors in the rat cerebral cortex, cerebellum, and spinal cord in vivo amounts to not less than 7.10 and not more than 7.33 pH units (Urbanics et al. 1978; Mutch and Hansen 1984; Siesjö et al. 1985; Kraig et al. 1983; Syková et al. 1988b; Syková and Svoboda 1990); pH_e differences in individual parts of the CNS could be due to differences in glia, to changes in the level of spontaneous neuronal activity, to the type and depth of anaesthesia, and to fluctuation of acid-base balance during artificial respiration. However, a number of studies have revealed that the pH_e in mammalian CNS is about 0.1–0.2 pH units more acid than the pH in arterial blood or CSF (Cragg et al. 1977; Javaheri et al. 1983; Kraig et al. 1983, 1986; Mutch and Hansen 1984; Siesjö et al. 1985; Syková and Svoboda 1990). The pCO_2 of nervous tissue and CSF are, in fact, equal; interstitial acidosis has also been found in in vitro preparations, e.g., in isolated turtle cerebellum (Nicholson et al. 1985; Chesler and Chan 1988), in lamprey brainstem (Chesler 1986, 1987), and in rat hippocampal slices (Balestrino and Somjen 1988).

The mechanism of interstitial acidosis is not clear. Classical acid extrusion mechanisms described in Sect. 6.3 may be responsible, although recent studies of cultured neurones and astrocytes suggest that glial cells – which have, for example, effective mechanisms for the extrusion of lactic acid (Walz and Mukerji 1988a,b) – may contribute to the generation of interstitial acid.

The pH_e base line in unstimulated spinal cord varies from 7.15 to 7.35 pH units (Fig. 57). Actual pH_e in the spinal cord was established by comparing the potential of the pH-ISM in CSF, and at various depths in the spinal cord, with that of a physiological solution poured onto the surface of the cord or placed in a small beaker next to the cord. The pH in supraspinal fluid measured just before the microelectrode penetrated the spinal cord was 7.35–7.40. Figure 57 shows the variations in pH_e base line in the lumbar spinal cord. In the upper

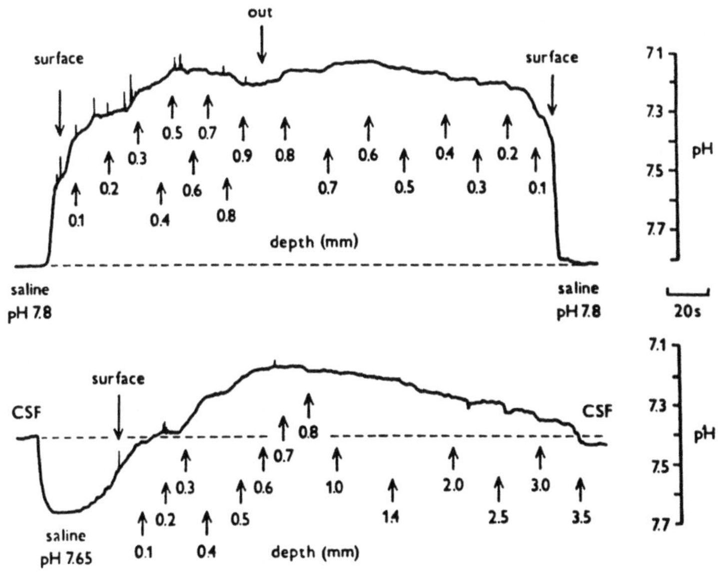

Fig. 57. pH_e at various depths in the unstimulated spinal cord of the rat. *Upper record*, the pH-ISM was introduced into the cord through saline, the pH of which was adjusted to pH 7.8 before it was poured over the dorsal surface of the cord. The *arrows* indicate the surface and various depths in the spinal dorsal horn. *Lower record*, the microelectrode was introduced into the cerebrospinal fluid (*CSF*) immediately after a small incision in the dura; saline, the pH of which was adjusted to pH 7.65, was then poured over the spinal cord before the electrode was introduced into the cord. The microelectrode penetrated the spinal cord until it reached the CSF on its ventral surface. (From Syková and Svoboda 1990)

dorsal horn and in the ventral horn the pH_e base line was only slightly more acid, at 7.25–7.35. The pH_e was most acid (about 7.15) in the lower dorsal horn at a depth of 500–700 μm from the dorsal surface. Variations in the pH_e base line had a similar depth profile to variations in the $[K^+]_e$ base line in unstimulated spinal cords. An alkaline shift in the pH_e base line of 0.05–0.10 pH units was found when synaptic activity was blocked by applying Mn^{2+} in a concentration of 4–5 mM onto the spinal cord surface. We may therefore assume that the pH_e resting level is more acid in the lower dorsal horn than in the upper dorsal horn, the ventral horn, and CSF, due to the release of H^+ or equivalent ions during a high level of spontaneous activity of spinal interneurones.

6.2.2 pH_e Changes Evoked by Stimulation of Afferent Input

In the brain, spinal cord, and nerves, the dynamic changes accompanying impulse activity have monophasic, biphasic, or triphasic time courses (for a review see Chesler 1990). The changes are alkaline-acid changes in which the

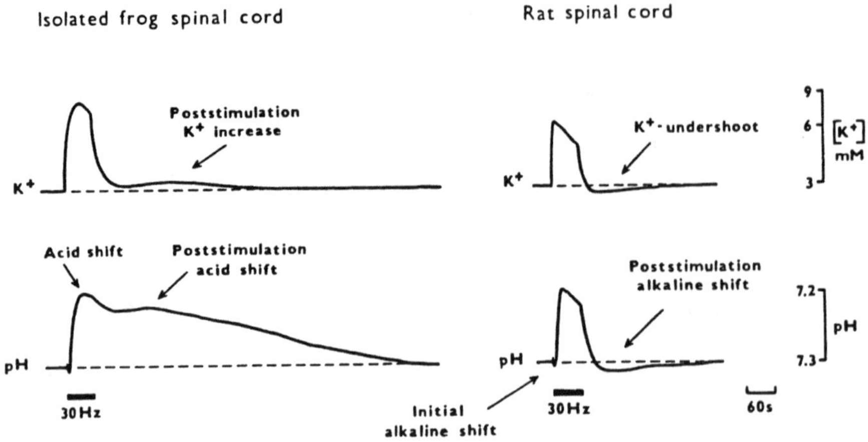

Fig. 58. Typical records of stimulation-related changes in pH_e and $[K^+]_e$ in the isolated frog spinal cord and in the rat spinal cord in vivo. Changes in pH_e and $[K^+]_e$ were produced in dorsal horns by electrical nerve stimulation at a frequency of 30 Hz applied for 60 s. Note the characteristic triphasic, alkaline-acid-acid changes in the isolated frog spinal cord, and alkaline-acid-alkaline changes in the rat spinal cord and their relation to the simultaneously recorded changes in $[K^+]_e$. (From Syková et al. 1988b)

acid shift is dominant (Fig. 58); they have been described in cerebellum (Kraig et al. 1983; Nicholson et al. 1985; Rice and Nicholson 1988), in cortex (Urbanic et al. 1978; Chesler and Kraig 1987), in hippocampus (Somjen 1984; Carlini and Ransom 1986), in spinal cord (Chvátal et al. 1988a, Syková et al. 1988a,b; Syková 1989, Syková and Svoboda 1990), in the vagus nerve (Endres et al. 1986), and in the optic nerve (Davis et al. 1987). Alkaline-acid changes in the ECS in association with neuronal activity have been described in the CNS during direct electrical stimulation of afferent input and during adequate stimulation. The pH_e evidently fluctuates by several tenths of a pH unit during stimulation. Changes in pH_e were greater in pathological states such as anoxia, ischaemia, and SD (see Sect. 6.6).

Repetitive electrical stimulation of nervous tissue or afferent input leads to depolarization of nerve cells and fibres, and to neuronal activity which is accompanied by ion shifts across the cell membranes and by typical dynamic changes in the pH_e. These depend on the intensity, frequency (see Fig. 64), and duration of stimulation. The greatest changes were observed in the cerebellum (Fig. 59; Kraig et al. 1983) and in the dorsal spinal horns, in response to repeated stimulation at a frequency of 100 Hz, at a stimulation intensity supramaximal for the stimulation of low-threshold afferent fibres (Fig. 60; Chvátal et al. 1988; Syková et al. 1988b; Syková and Svoboda 1990). During repetitive stimulation in the spinal cord the acid change attains a plateau in 15–20 s: both in the rat spinal cord in vivo (Figs. 58 and 60), and in the frog spinal cord in vitro (Figs. 58 and 61). The pH_e acid shifts have a slower time course than simultaneously recorded $[K^+]_e$ changes (e.g., in the spinal cord the

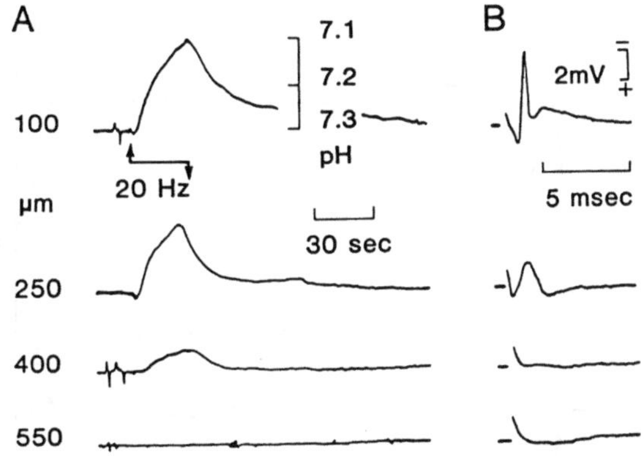

Fig. 59A,B. Depth profile of pH_e changes associated with repetitive surface stimulation in the rat cerebellum. **A** Depth profile of pH_e changes during surface stimulation for 30 s at 20 Hz along a parallel fibre; the beam starts with *upward arrow* and stops with *downward arrow*. Stimulus was repeated at increments of 150 μm as the pH-ISM was advanced vertically through the folium. Depth was corroborated by comparison with typical laminar changes in field potentials shown in **B**. (From Kraig et al. 1983)

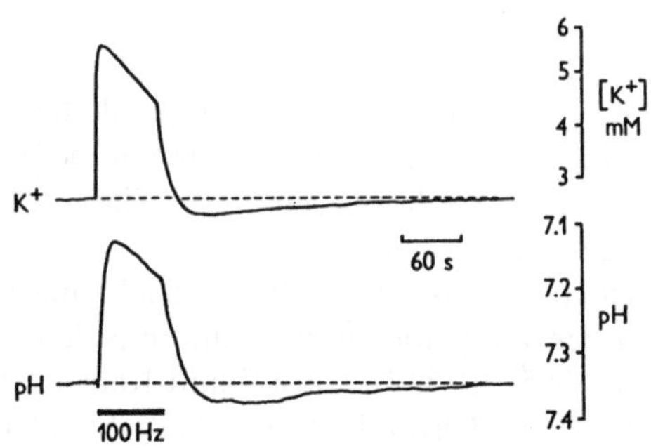

Fig. 60. $[K^+]_e$ and pH_e changes in the dorsal horn of segment L_4 of the adult rat spinal cord. Repetitive bipolar electrical stimulation through a pair of needles (100 Hz, 60 s) inserted into the planar muscles of the ipsilateral hind paw. (From Syková and Svoboda 1990)

$[K^+]_e$ increment reaches the maximum in only 6–9 s). If stimulation is continued, i.e., after the maximum acid change has been reached, the acid value gradually returns to the original pH_e (an alkaline shift occurs), as is the case in $[K^+]_e$ changes. As previously mentioned in Sect. 4.1.1, the drop in $[K^+]_e$ which follows the initial increase is primarily due to activation of the Na^+/K^+ pump. When stimulation is discontinued, K^+, Ca^{2+}, and H^+ in the ECS all follow the same time course in their return to the original level (Fig. 61).

Although the acid change usually predominates during stimulation, there is a distinct biphasic or triphasic course to the pH_o changes. The acid change is often preceded by a small initial alkaline shift (about 0.05–0.1 of a pH unit) lasting 2–3 s (Fig. 60). Its latency is the same as for the onset of the increase in

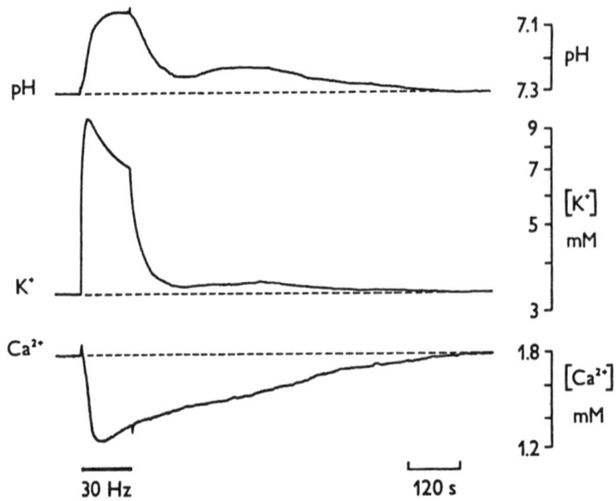

Fig. 61. pH_e, $[K^+]_e$, and $[Ca^{2+}]_e$ changes in the dorsal horns of the isolated frog spinal cord. Repetitive supramaximal stimulation of dorsal root VIII (30 Hz, 120 s). The recordings were made with three double-barrelled ISMs introduced to the same depth (350 μm), with their tips 100–200 μm apart. (From Chvátal et al. 1988)

$[K^+]_e$. Directly following the alkaline shift is the acid change, which lags somewhat behind the beginning of the $[K^+]_e$ increases. On repetitive stimulation in the dorsal horns and cerebellar cortex, this acid change amounts to 0.15–0.25 of a pH unit.

The initial alkaline shift is partially masked by later acidification. It is generally larger in aquatic animals and in isolated preparations (Chesler 1985; Endres et al. 1985, 1986a,b; Carlini and Ransom 1986; Chesler and Chan 1988; Davis et al. 1987; Chvátal et al. 1988; Syková 1989; Jendelová and Syková 1990), although it dominates the acid shift in the spinal cord of newborn rats, where the gliogenesis is not yet complete (Jendelová and Syková 1990).

In the spinal cord of the rat, during the first eight postnatal days, stimulation of afferent input evoked large alkaline changes in the extracellular microenvironment, followed by slow, small acid shifts after stimulation had been discontinued. These changes progressively decreased as the rat developed, and about 10 days after birth, the acid shift became the dominating pH_e change – at the same time as the K^+ ceiling level started to decrease (see Fig. 76; Jendelová and Syková 1990). Alkaline shifts evoked by stimulation have been also found in the unmyelinated optic nerve of 1-day-old rats, the acid shift occurring at about 10 days or later, when $[K^+]_e$ changes were no longer so great (Ransom et al. 1985b). In the rat spinal cord, as well as in the optic nerve, the proliferation of astrocytes and oligodendrocytes, and myelination occurs postnatally (Gilmore 1971; Gilmore et al. 1982; Yamate and Ransom 1985; Sims et al. 1985; Foster et al. 1982) and their time course correlates well with the observed development of the acid shift and with the decrease in the K^+ ceiling level (Ransom et al. 1985b).

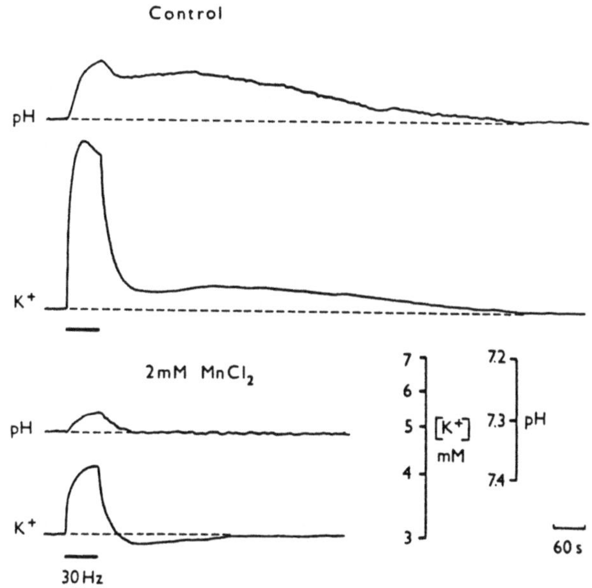

Fig. 62. Effect of a block of synaptic transmission, produced by adding 2 mmol/l $MnCl_2$ to the Ringer solution, on pH_e and $[K^+]_e$ changes induced in the dorsal horns of the isolated frog spinal cord by stimulation of dorsal root VIII (30 Hz, 60 s). (Syková, Chvatâl and Jendelová, unpublished figure)

Enhancement of the initial alkaline shift has been also observed in in vitro preparations perfused with a solution buffered with 15 mmol/l of HEPES instead of 35 mmol/l HCO_3^- (Chesler and Chan 1988), i.e., in solutions with lower buffering power. Moreover, Krishtal et al. (1987) recorded an alkaline shift with a latency of 20 ms in hippocampal slices when they used optic techniques which recorded a shorter response time than that possible using of pH-ISMs. Rice and Nicholson (1988) also found large initial alkaline shifts in vivo in skate cerebellum, where $[HCO_3^-]$ is relatively low.

When stimulation is discontinued, the pH_e returns to the original resting level. The course of the decrease in pH_e, like that of other ionic changes in the ECS, varies according to the animal studied and the preparation. For instance, different courses of redistribution have been observed in the isolated frog, mouse, and golden hamster spinal cord, and in the rat spinal cord in vivo. A triphasic (alkaline-acid-alkaline) course for the pH_e (Fig. 60) is typical of the mammalian spinal cord in vivo, whereas a biphasic acid course is often seen in the frog spinal cord in vitro (Fig. 61). In the latter, the first wave (an acid shift) is formed during stimulation, followed by a second acid shift which does not appear until stimulation has been terminated. This second acid wave is the result of an increase in spontaneous activity in the dorsal horns after tetanic stimulation and – like the late increase in $[K^+]_e$ and the late depolarization of primary afferents – it disappears when synaptic activity is blocked (Fig. 62). We therefore have, in principle, a variable rate according to which the tissues

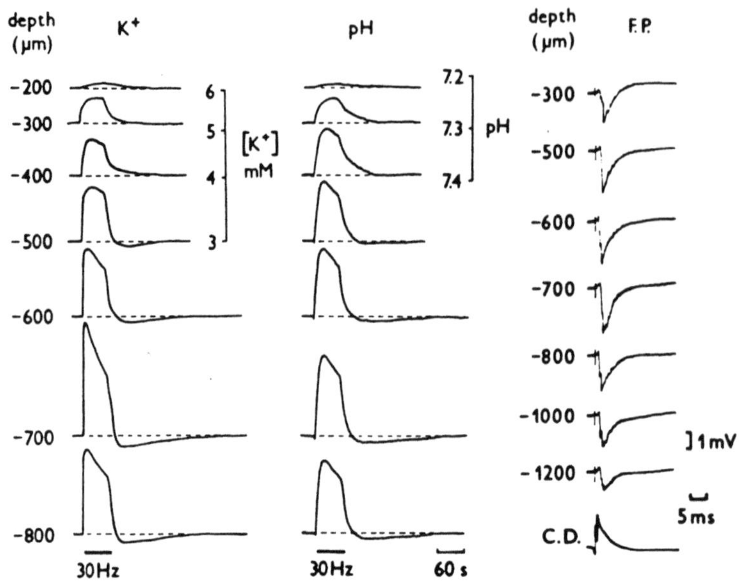

Fig. 63. A depth profile of the stimulation-evoked changes in the extracellular K^+, pH, and focal potentials (*F.P.*). Increase in $[K^+]_e$ and alkaline-acid-alkaline changes in pH_e were evoked by repetitive electrical stimulation at a frequency of 30 Hz applied for 60 s through a pair of needles inserted into the planta. Focal potentials (*F.P.*), recorded by the reference barrel of a pH-ISM, and the potential from cord dorsum (*C.D.*) were evoked by single electrical stimuli. The K^+-ISM was inserted into the same area after withdrawal of the pH-ISM. (From Syková and Svoboda 1990)

of different experimental animals cope with ionic changes in the ECS (including changes in the pH_e) accompanying neuronal activity. After repetitive stimulation, amphibians always displayed a longer increase in excitability than mammals. In rats, in vivo, the redistribution of accumulated K^+, the decrease in $[Ca^{2+}]_e$, and acid-base changes are balanced relatively quickly – even without a block of synaptic transmission; after stimulation has been discontinued we see a rapid drop in $[K^+]_e$ and pH_e changes, and always find an undershoot below the resting levels. The mechanisms of the individual phases of pH_e changes will be discussed in Sect. 6.3.

During stimulation of hind limb peripheral nerves or dorsal roots, we found that pH_e changes in the spinal cord of both mammals and amphibians had a characteristic depth profile. It was identical to the depth profile for $[K^+]_e$ and $[Ca^{2+}]_e$ changes evoked by stimulation, i.e., maximum changes were observed in the deeper layers of the dorsal horns (Fig. 63). At the site of the greatest pH_e changes – i.e., an initial alkaline shift, an acid change, and a post-stimulation alkaline undershoot – we also found the largest focal potentials, evoked by peripheral nerve stimulation with a single electric pulse and recorded with an ISM reference channel (Fig. 63). The greatest pH_e changes are, therefore, to be found in Rexed's laminae III–V (Rexed 1954) in the rat spinal

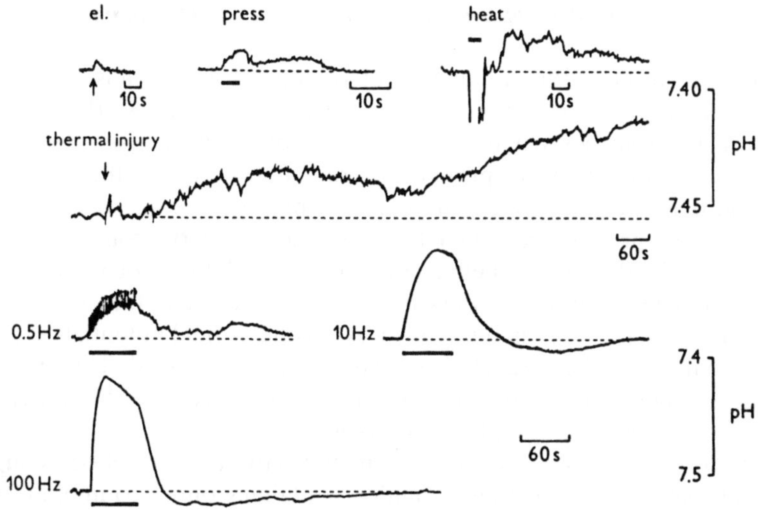

Fig. 64. Transient decrease in pH_e in the L_4 spinal segment in response to a single electrical stimulus applied through a pair of acupuncture needles inserted into the planta (*el.*), to the application of pressure to the toes (*press*), to the application of about 1 ml of hot water (*heat*), or of 2 ml of hot oil onto the hind paw (*thermal injury*), and in response to repetitive electrical stimulation at frequencies 0.5, 10, and 100 Hz. *Horizontal bars* indicate stimulus duration. Records at a depth of 600–700 μm. (From Syková and Svoboda 1990)

cord, and in Ebbesson's dorsal and lateral fields (Ebbesson 1976) in the frog spinal cord, i.e., the maximum changes are localized in the region where low-threshold afferent fibres terminate on spinal interneurones, and where neurone density per area unit is the greatest (Aitken and Bridger 1961).

6.2.2.1 pH_e Changes Evoked by Adequate Stimulation of Skin Nociceptors

Like $[K^+]_e$ changes, pH_e changes can also be induced in the dorsal horns by adequate stimulation of skin receptors. Acid shifts of just 0.01–0.05 of a pH unit have been found in rat dorsal horns at a depth of 400–700 μm, i.e., in Rexed's laminae III–V (Syková et al. 1988b; Syková and Svoboda 1990). The smallest changes were produced by tactile stimulation of the skin of the hind limb or by slight pressure on the skin; these changes lasted 10–20 s. Greater pH_e changes, lasting up to 30 s, were evoked by nociceptive stimuli, such as pressing a toe with forceps, or applying radiant heat or hot water to the skin of a hind paw (Fig. 64). Local skin injury, such as scalding a paw with hot water or oil, or the intradermal injection of formalin or terpentine, produced an inflammatory reaction. This type of injury was accompanied by a long acid shift of 0.05–0.1 of a pH unit in the dorsal horns; the pH_e changes after injury thus accompany an increase in $[K^+]_e$, as described in Sect. 4.1.4.

6.2.3 *Effect of Block of Synaptic Transmission on pH_e Changes*

Repetitive electrical stimulation of peripheral input leads to accumulation of K^+ and to other dynamic ionic changes in the ECS; these changes result largely, from postsynaptic neuronal activity, but are (see Sect. 4.1) also the outcome of activity in primary afferent fibres. It can, therefore, be assumed that dynamic pH_e changes are partly of postsynaptic (i.e., neuronal) origin and partly of presynaptic origin (i.e., due to depolarization of primary afferent fibres). In the rat cerebellum in vivo, a Mn^{2+} block of synaptic transmission from parallel fibres to Purkinje cells counteracted the initial alkaline shift (Kraig et al. 1983); an analysis of extracellular field potentials showed that the excitability of the parallel fibres had not been depressed. This and other studies in the spinal cord (Jendelová and Syková 1991) suggest that the initial alkaline shift is a result of synaptic transmission.

In the isolated frog spinal cord, synaptic activity was blocked by superperfusing the spinal cord with Ringer solution containing only 0.5 mmol/l $CaCl_2$ together with either 2–4 mmol/l $MnCl_2$ or 20 mmol $MgCl_2$. The initial alkaline shift disappeared (Syková 1989), the acid pH_e changes evoked by repetitive stimulation fell to 25%–35%, and spontaneous activity and the second post-stimulation acid shift also disappeared (Fig. 62). These results show that depolarized postsynaptic elements are responsible for the majority of changes in pH_e. Like other dynamic extracellular ionic changes, however, some of the pH_e changes may result from depolarization of the primary afferent fibres themselves.

6.2.4 *pH_e Changes Induced by K^+ Depolarization*

It can be seen from most of the figures in this chapter that pH_e changes are always accompanied by changes in $[K^+]_e$. It is unlikely that acid changes are the outcome of a decrease in $[SID]_e$ resulting from a change in the concentration of $[K^+]_e$ and other strong ions in the ECS. Computations based on the measurement of concentration changes in all the strong ions by means of ISMs actually show that repetitive stimulation is attended by an increase, not a decrease, in $[SID]_e$, which should explain the acid changes observed in the pH_e (Kraig et al. 1983). An increase in $[SID]_e$ is associated with an alkaline shift, which has been observed in the cerebellar, cortex, and at the onset of stimulation in most of our experiments in the spinal cord (Figs. 60, 63, 65; Kraig et al. 1983; Syková et al. 1988b; Chvátal et al. 1988; Syková and Svoboda 1990). Perfusion of the rat cerebellum in vivo (Kraig et al. 1983) or the isolated frog spinal cord (Chvátal et al. 1988) with Ringer solution with a raised K^+ concentration enhanced the initial alkaline shift (Fig. 65).

The increase in K^+ in the perfusion solution depolarizes the neuronal elements and causes the pH_e to fall. As seen in the rat cerebellum (Kraig et al. 1983), a 2–6 mmol/l increase in $[K^+]_e$ in the deeper layers of the dorsal horns produces an acid shift of 0.2–0.6 of a pH unit (Fig. 65; Chvátal et al. 1988). Repetitive electrical stimulation under conditions of K^+ depolarization and a raised

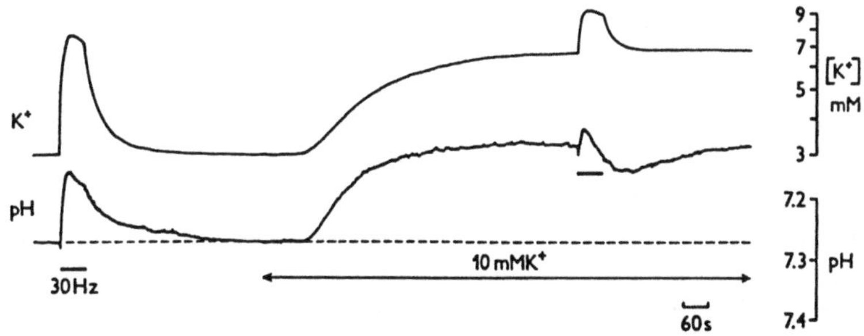

Fig. 65. The decrease in the pH_e (an acid shift) in the isolated frog spinal cord after raising $[K^+]_e$ in the Ringer solution from 3 to 10 mmol/l K^+. The application time is marked by the *arrows*. Dynamic pH_e and $[K^+]_e$ changes were evoked by electrical stimulation (30 Hz, 60 s) of a dorsal root before and after adding 10 mmol/l K^+. (From Chvátal et al. 1988)

$[K^+]_e$ led to smaller acid changes in the pH_e, but to bigger (initial and post-stimulation) alkaline shifts. Reduction of $[K^+]_e$ from 3 to 1.5 mmol/l, which hyperpolarized the neuronal membrane, led to an alkaline shift of about 0.5 of a pH unit (Chvátal et al. 1988). Tetanic stimulation of afferent input in the presence of a low $[K^+]_e$ produced a greater change in both $[K^+]_e$ and pH_e than occurred in normal Ringer solution. These findings are in agreement with the experiments of Thomas (1988a,c) on isolated snail neurones, in which depolarization of the membrane evoked acid surface changes and hyperpolarization produced alkaline surface changes. Thomas suggested that these changes were caused by the opening of specific H^+ channels. However, in studies where various specific inhibitors of membrane transport processes were used (as listed and described in Sect. 6.3), none of the inhibitors influenced the K^+-evoked pH_e acid shifts. On the other hand, activity-dependent pH_e transients are accompanied by rapid changes in glial intracellular pH. The glial pH_i is sensitive to membrane potential, and is therefore affected by the release of K^+ from active neurones (Chesler and Kraig 1987, 1989) or by the increase of $[K^+]_e$ in the perfusion solution (see Sect. 6.4).

6.3 Mechanisms of pH_e Changes in the CNS

Substances which inhibit synaptic transmission, metabolic processes, membrane ion transport systems, enzymatic reactions, and ion channels are used for studying the mechanisms of pH_e shifts. Specific inhibitors were initially used in studies of pH_i changes, but their use has recently been extended to the study pH_e. This chapter describes the effect of these inhibitors on different phases of pH_e changes induced by tetanic electrical stimulation of afferent input.

6.3.1 Effect of Sodium Fluoride

Sodium fluoride, which blocks the glycolytic part of metabolic processes and lactate production, did not have a significant effect on pH_e changes induced by tetanic stimulation of peripheral input (Syková et al. 1988b). This suggests that the lactate production demonstrated under the extreme conditions summarized in Sect. 6.6 did not play an important role in electrical and adequate stimulation. Siesjö et al. (1985) showed that even in an epileptic seizure, the lactic acid level in the ECS did not change over a period of 30–60 s. These authors suggested that Na^+/H^+ changes participated in the acid changes observed in the first phases of raised activity. Another argument against the role of lactate in the mechanism of pH_e changes during neuronal activity is that it takes the raised lactate level about 15 min to return to normal during epilepsy and ischaemia, whereas after electrical stimulation of afferent input the pH_e returns to normal at the same rate as the change in $[K^+]_e$, i.e., in only 30–60 s in mammals. In peripheral nerve stimulation lasting 5 min or more, pH_e changes were no greater than in short stimulation (30 s); on the contrary, as soon as the maximum acid change had been attained, an alkaline shift to the original level occurred (e.g., in Fig. 60). As regards the accumulation of lactate, it can be assumed that the return to the resting level is proportional to the length of stimulation and to the size of lactate production. The findings to date thus indicate that lactate production does not play an important role in the mechanism of dynamic pH_e changes after electrical and adequate stimulation.

6.3.2 Effect of Ouabain

H^+ can also be extruded by means of the Na^+/K^+ pump, whereby H^+ is expelled in exchange for K^+ (Fig. 56i). The Na^+/K^+ pump plays an important role in the clearance of K^+ which accumulates in the ECS during the stimulation of neuronal elements and depolarizes the neuronal membrane. The effect of the specific Na^+/K^+-ATPase inhibitor ouabain on the resting $[K^+]_e$ and pH_e level and on changes evoked by stimulation has been the subject of several studies (Kraig et al. 1983; Chvátal et al. 1988; Syková and Svoboda 1990). It has been found that, after the administration of ouabain, the resting $[K^+]_e$ rises and primary afferent fibres are depolarized; changes evoked in $[K^+]_e$ by stimulation are greater and their redistribution takes longer (see Fig. 5). The resting pH_e, on the other hand, falls by 0.05–0.07 of a pH unit (see Fig. 67); the ceiling pH_e level in mammals, i.e., the maximum acid shift, however, was not much lower – it decreased by 15.3 ± 2% (Syková and Svoboda 1990). In mammals (Fig. 66) and amphibians (Fig. 67), ouabain slows down acid changes in the ECS during stimulation. Perfusion of the isolated frog spinal cord with ouabain in 10^{-5} mol/l concentration reduced acid changes evoked by stimulation by 20%–50%, while the $[K^+]_e$ ceiling level increased. These results support the hypothesis that ouabain blocks the reabsorption of K^+ (probably into both neurones and glial cells) and the extrusion of H^+ by the mechanism

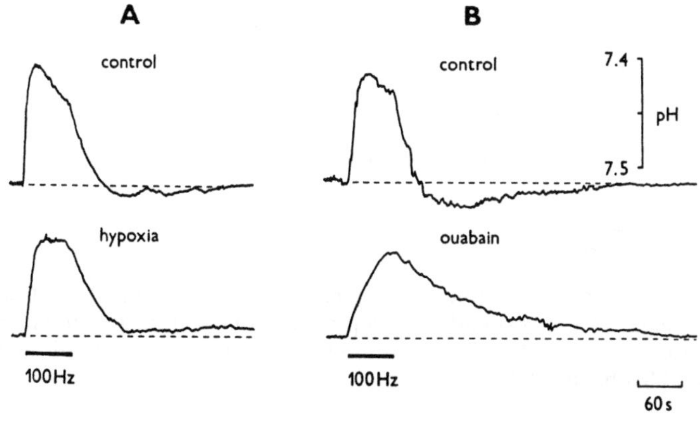

Fig. 66A,B. Depression of pH_e changes evoked by repetitive electrical stimulation (100 Hz, 60 s) after partial respiratory arrest when blood pressure decreased to 70 mm Hg (hypoxia) (**A**), and after application of 10^{-4} mol/l ouabain onto the surface of the spinal cord (**B**). The data are from two animals. (From Syková and Svoboda 1990)

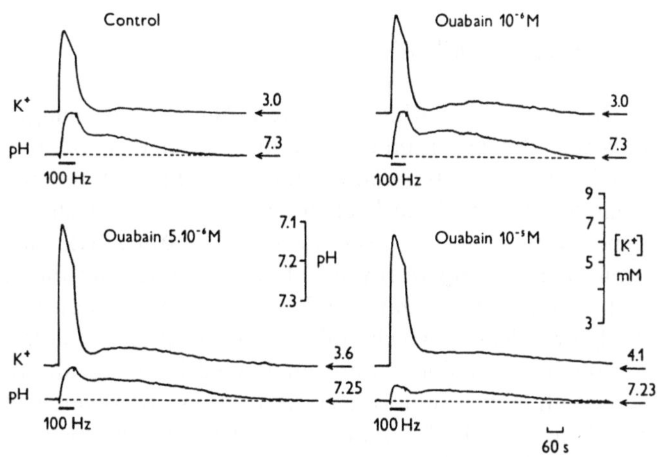

Fig. 67. Effect of ouabain in 10^{-6}, 5.10^{-6}, and 10^{-5} mol/l concentration on the resting $[K^+]_e$ and resting pH_e in the isolated frog spinal cord and on changes evoked by dorsal root stimulation (100 Hz, 60 s). The *numbers* over the *arrows* denote the resting level values in mmol/l and in pH units, and changes in these values after 20 min exposure to ouabain. (From Chvátal et al. 1988)

illustrated in Fig. 56i. In addition, ouabain or hypoxia block both the K^+ undershoot below the resting level after electrical stimulation has terminated and the post-stimulation alkaline shift (Fig. 66); it also prolongs the return of both these ionic changes to normal pre-stimulation values (Fig. 66; Chvátal et al. 1988; Syková et al. 1988b).

These findings indicate that activation of the Na^+/K^+ pump may be an important mechanism in pH_e homeostasis. It is, however, possible that a ouabain

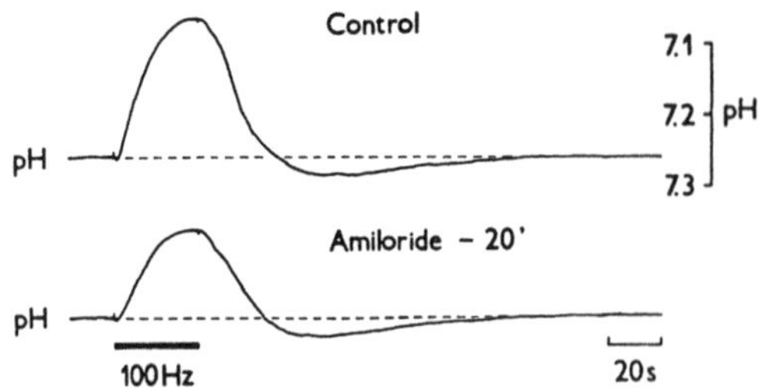

Fig. 68. Effect of amiloride applied onto the rat spinal cord surface. Records show changes in pH_e evoked at a depth of about 650 μm by stimulation at 100 Hz, 30 min prior to (*control*), and 20 min after application of 10^{-3} mol/l amiloride. (From Syková and Svoboda 1990)

block of the Na^+/K^+ pump depresses the membrane Na^+ gradient which energy is utilized by Na^+/H^+ exchange and associated $Na^+/H^+/HCO_3^-/Cl^-$ exchange, with resultant secondary inhibition of acid changes.

6.3.3 *Effect of Amiloride*

In concentrations which blocked Na^+/H^+ exchange (10^{-4}–10^{-3} mmol/l), amiloride did not produce any marked changes in the resting pH_e. During tetanic electrical stimulation, the acid shifts in the rat spinal cord and the isolated frog spinal cord fell by 5%–10% (10^{-4}) and 20%–30% (10^{-3} mol/l) respectively (Fig. 68; Syková et al. 1988b; Syková and Svoboda 1990). Amiloride had practically no effect on either the initial alkaline shift or the post-stimulation alkaline undershoot. Simultaneously recorded changes in $[K^+]_e$ or DRPs – which reflect any effect of an administered substance on synaptic transmission – were unaffected by amiloride in 10^{-4} mol/l concentration. With higher amiloride concentrations, however, neither the resting $[K^+]_e$, nor changes in $[K^+]_e$ could be measured exactly, owing to amiloride sensitivity of the K^+-ISMs (Syková and Svoboda 1990).

These findings show that amiloride-sensitive Na^+/H^+ exchange is one of the mechanisms which participates in the release of excess H^+ from activated nerve cells or glial cells in the vertebrate CNS. However, since amiloride does not block more than 20%–30% of acid pH_e changes occurring during repetitive stimulation, Na^+/H^+ exchange is obviously neither the only nor the most important mechanism.

6.3.4 *Effect of SITS and DIDS*

In the rat spinal cord the blockade of Cl^-/HCO_3^- exchange and/or $Na^+/H^+/HCO_3^-/Cl^-$ antiport by the stilbene derivatives SITS and DIDS resulted in an acid shift in the pH_e base line of 0.01–0.05 pH units (Syková and Svoboda

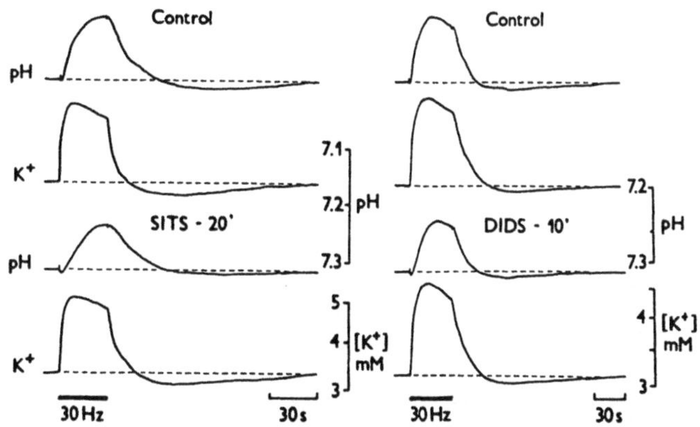

Fig. 69. Effects of SITS and DIDS applied onto the dorsal surface of the rat spinal cord. Simultaneously recorded changes in $[K^+]_e$ and pH_e in the L_4 spinal cord segment at a depth of about 700 μm were evoked by repetitive electrical stimulation at a frequency of 30 Hz applied through a pair of needles inserted into the planta. Note that the initial alkaline shift was enhanced, the acid shift was depressed and its rise time was slowed down. Records were taken 20 min after application of 3×10^{-3} mmol/l SITS and 10 min after application of 2×10^{-3} DIDS. The simultaneously recorded increase in $[K^+]_e$ was no different after the application of SITS and DIDS. The data are from two animals. (From Syková and Svoboda 1990)

1990). The stimulation-evoked acid shifts were depressed by 25%–30% after the application of SITS (1–3 mmol/l) or DIDS (1–2 mmol/l) to the spinal cord surface. Moreover, the rise time of the acid shift was slowed down and the initial alkaline shift was enhanced (Fig. 69). The application of SITS and DIDS in these experiments affected neither the $[K^+]_e$ base line, nor the concomitantly recorded stimulation-evoked transient increase in $[K^+]_e$ (Fig. 69). Their application in higher concentrations further depressed pH_e, and also depressed $[K^+]_e$ transients, suggesting that non-specific effects are involved. Evidently, up to 30% of the block of acid changes may, therefore, be the outcome of a block of Cl^-/HCO_3^- exchange, $Na^+/H^+/HCO_3^-/Cl^-$ exchange, or Na^+-HCO_3^- cotransport in either neurones or glia. Our findings provide evidence that bicarbonate and chloride shifts which can be blocked by stilbene derivatives play an important role in pH_e and pH_i regulation.

6.3.5 Effect of Acetazolamide

In the CNS, carbonic anhydrase (CA), an enzyme catalysing the hydration of CO_2 to H_2CO_3, which is dissociated to H^+ and HCO_3^- (see Sect. 6.1), is found chiefly in glial cells and myelinated nerve fibres (Giacobini 1962; Kimelberg et al. 1978; Cammer et al. 1985; Roussel et al. 1979). The rate of the reaction described above, or of the reaction of dissociated CO_2 and OH^- (H_2O) to HCO_3^-, increases in order from several tenths of a second to a few micro-

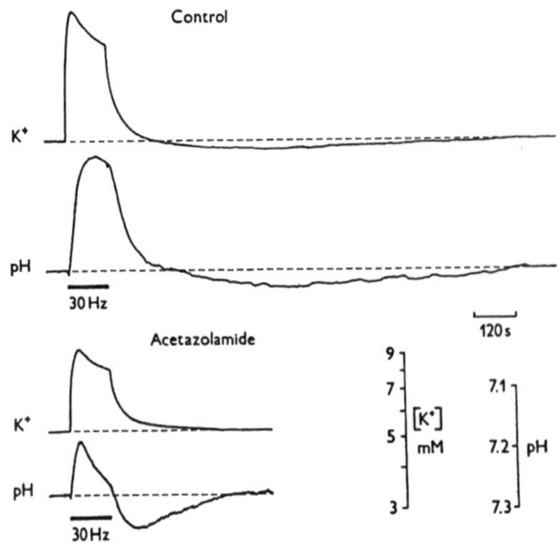

Fig. 70. Effect of acetazolamide (4 mmol/l) on simultaneously recorded changes in pH_e and $[K^+]_e$ in the frog dorsal horn, evoked by tetanic stimulation of dorsal root at 30 Hz. Note the post-stimulation alkaline shifts. (From Chvátal et al. 1988)

seconds (Maren 1967; Stewart 1981). Since the main buffer system in the extracellular fluid in vertebrates is the HCO_3^--CO_2 system, a block of CA probably has a pronounced effect on pH_e homeostasis. As observed in the cerebellar cortex by Kraig et al. (1983), the application of acetazolamide, a non-competitive CA inhibitor, caused an alkaline shift of about 0.05 of a pH unit in the resting pH_e in the spinal cord.

In a concentration of 10^{-6}–10^{-3} mmol/l, acetazolamide substantially accelerated the onset of stimulation-evoked acid changes. Acid pH_e changes after application of acetazolamide were as fast as, or even faster than, changes in $[K^+]_e$, in both the isolated frog spinal cord and the rat spinal cord. A possible explanation for these findings is that some CA is also localized extracellularly where the buffering power of the extracellular fluid is decreased. Acetazolamide did not, however, affect stimulation-induced changes in $[K^+]_e$, Kraig et al. (1983) showed that oscillations of the pH_e produced in the cerebellum by the administration of K^+ were greater in the case of CA inhibition.

When applied in higher concentrations (from 5.10^{-3}), acetazolamide did, however, reduce stimulation-evoked acid shifts in the isolated frog spinal cord (Fig. 70) while enhancing the initial alkaline shift (Chvátal et al. 1988). These effects were even more prominent in experiments on the isolated frog spinal cord, where the substitution of HEPES for bicarbonate buffer in the Ringer solution led to depression of acid changes and to enhancement of the initial alkaline shift to about 0.2 of a pH unit (Fig. 71; Syková 1989). The effects of acetazolamide were similar in the isolated spinal cord of immature rats (2–4 days old). In these immature animals gliogenesis is not yet complete and stimulation of the dorsal root produces an alkaline shift of about 0.2 pH units (Jendelová and Syková 1990). The typical acid shift evoked by afferent stimula-

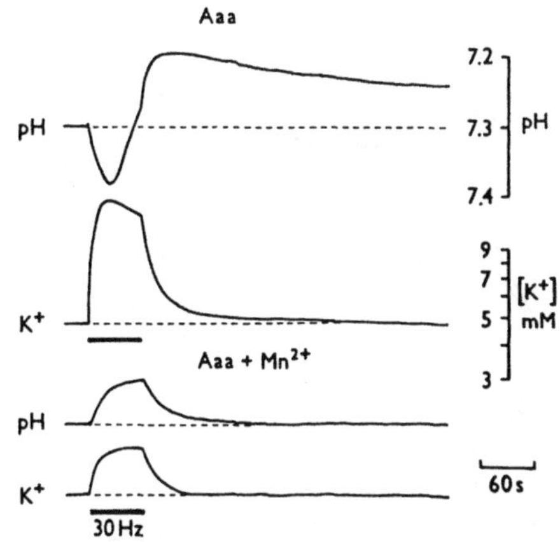

Fig. 71. Effect of acetazolamide (*Aaa*) in 10^{-4} mol/l concentration on simultaneously recorded changes in pH_e and $[K^+]_e$ in the isolated frog spinal cord (superfused with Ringer solution buffered with HEPES), evoked by tetanic stimulation of dorsal root at 30 Hz. Note the enhanced initial alkaline shift and the onset of acid shift before the end of stimulus train. *Lower* records show the effect of Mn^{2+} (2 mmol/l) added to the Ringer solution containing acetazolamide on the pH_e and $[K^+]_e$ changes. Note the complete block of the initial alkaline shift, the decrease in the stimulation-evoked acid shift, and the decrease in $[K^+]_e$. (From Syková 1989)

tion is seen in animals 10–12 days old; in these animals acetazolamide does not influence the alkaline shifts, but the acid shifts are depressed by about 80%.
It is, therefore, evident that CA significantly enhances the buffering capacity of the microenvironment in the CNS in, it is reasonable to assume, both the extracellular fluid and the glial cells.

6.3.6 Effect of Furosemide

In the CNS, some of the K^+ released from neuronal elements into the ECS is removed by glial cells by the following mechanisms: (a) the spatial buffer system, (b) active reabsorption of K^+ utilizing Na^+/K^+-ATPase, and (c) K^+-Cl^- cotransport or $Na^+/K^+/Cl^-$ cotransport (see Sect. 3.3.2). K^+-Cl^- cotransport which causes enlargement of the glial cells due to its association with H_2O transfer (Fig. 72A), can be blocked with furosemide (Kimelberg and Frangakis 1985; Walz and Hertz 1982, 1983; Tas et al. (1986).
We studied the effect of furosemide on the resting pH_e and on changes induced in pH_e and $[K^+]_e$ by stimulation. Furosemide causes the resting pH_e to shift to the alkaline side by 0.1–0.15 of a pH unit, but raises the resting K^+ level by several tenths of a mmol/l. It blocks acid pH_e changes evoked by stimulation in the rat spinal cord and the isolated frog spinal cord by 20%–30% (Fig. 72B; Syková et al. 1988b), and also blocks stimulation-induced $[K^+]_e$ changes by 10%–20%.

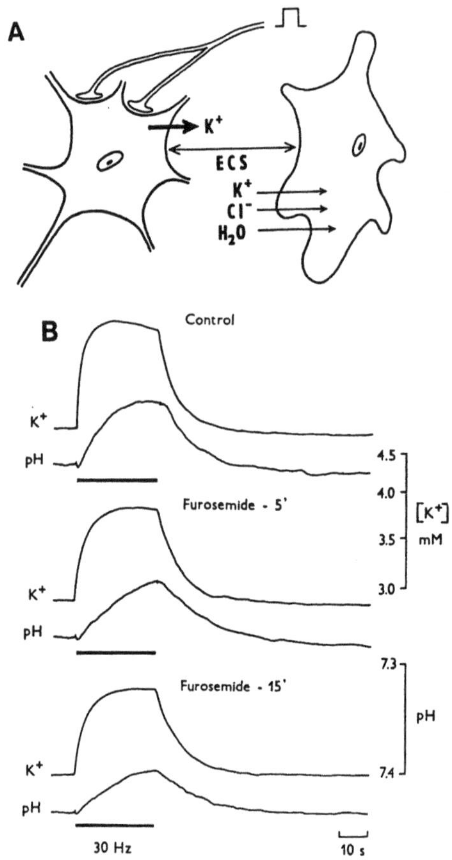

Fig. 72. A Schematic illustration of K^+-Cl^- cotransport. K^+ released from activated neurones stimulates cotransport of K^+-Cl^- in the glial cells. This is associated with a water influx which leads to enlargement of the glial cells and to reduction of the size of the ECS. **B** Effect of furosemide (10^{-3} mol/l) on dynamic pH_e and $[K^+]_e$ changes induced in segment L_4 of the rat spinal cord by repetitive electrical stimulation of a hind paw. Furosemide was applied to the dorsal surface of the spinal cord for 5 min and for 15 min. (Syková and Svoboda, unpublished)

The extent to which furosemide affects stimulation-induced changes in the size of the ECS is not yet clear (see Chap. 7). However, furosemide does affect Na^+, Cl^-, H^+, and HCO_3^- transport processes across glial cell and neuronal membranes, and thereby influences pH_e homeostasis.

6.3.7 Effect of Block of H^+ Channels

Various divalent and trivalent cations were used to test the possibility that H^+ or its equivalents leave or enter cells during repetitive electrical stimulation via voltage sensitive H^+ channels. The application of Mn^{2+} or Mg^{2+} to the surface

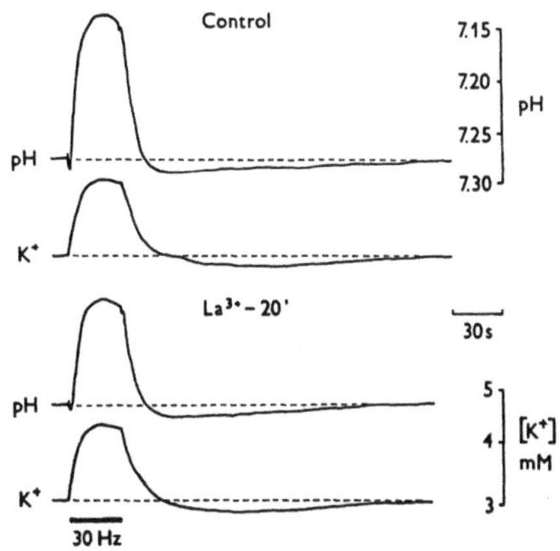

Fig. 73. Changes in $[K^+]_e$ and pH_e in the L_4 spinal segment – evoked by electrical stimulation (30 Hz, 30 min) – recorded at a depth of 650 μm prior to (*control*) and 20 min after application of 3×10^{-3} La^{3+} onto the rat spinal cord surface. Note the depression of the initial alkaline shift in addition to the acid shift. (From Syková and Svoboda 1990)

of the spinal cord depressed both pH_e and $[K^+]_e$ transients, owing to the blockade of synaptic transmission (see Fig. 62). The application of Cd^{2+} or Co^{2+} also blocked synaptic transmission, but the application of La^{3+} did not have this effect. In addition, we found that the sensitivity of pH-ISMs was dramatically reduced by Cd^{2+}, Co^{2+}, or Zn^{2+} in concentrations of 10^{-6}–10^{-3} mol/l (Syková 1989; Syková and Svoboda 1990), but not by similar La^{3+} concentrations. All these ions were shown to block voltage-sensitive H^+ channels in snail neurones (Meech and Thomas 1987; Thomas 1988b); however, we could only test the effect of La^{3+} on stimulation-evoked transient pH_e changes.

The application of $LaCl_3$ in a concentration of 1 mmol/l to the surface of the rat spinal cord, or perfusion of the isolated frog spinal cord with Ringer solution containing $LaCl_3$, lowered the resting pH_e by 0.05–0.3 pH units. In rats, La^{3+} depressed the stimulation-evoked acid shift by 23.6 ± 5.2%, although the time course of the shift was not changed; it depressed the stimulation-evoked initial alkaline shift by more than 50% (Fig. 73). The post-stimulation alkaline undershoot was not altered by the application of La^{3+} and there was no effect on stimulation-evoked K^+ transients (Syková and Svoboda 1990). In the isolated frog spinal cord La^{3+} depressed the initial alkaline shift which can be enhanced in HEPES buffer and after the application of acetazolamide by about 60%–80% (Jendelová et al. 1990).

These and other results suggest that the initial activity-related extracellular alkaline shifts may result from the influx of H^+ via voltage sensitive H^+ channels in activated neurones (Chesler and Chan 1988; Syková and Svoboda 1990).

The results of studies using specific inhibitors can be summarized as follows:

Fig. 74. Intracellular pH transients in rat cortical astrocytes. Cells were impaled with double-barrelled pH microelectrodes in vivo. Electrical stimulation of cortical surface caused rapid intracellular alkaline shift which could be correlated with the time course of membrane depolarization. Note the progressive acid shift after delivery of repetitive trains (20 Hz). (From Chesler and Kraig 1989)

1. The initial alkaline shift has a different mechanism from that of the acid shift and the post-stimulation alkaline undershoot. The initial alkaline shift is significantly blocked by La^{3+} and its short time course and the effective blockade by Mn^{2+} and La^{3+} suggest that the influx of H^+ via voltage-sensitive H^+ channels could be the dominating mechanism. Since it was abolished by Mn^{2+} (a Ca^{2+} channel blocker), the initial alkaline shift must be directly associated with Ca^{2+}-mediated electrical activity of the neurones; it also dominates over the acid shift in immature nervous tissue where gliogenesis is not yet complete.
2. The acid shift which develops during stimulation has at least four mechanisms: (a) Na^+/H^+ exchange, (b) Cl^-/HCO_3^- exchange or $Na^+/Cl^-/H^+/HCO_3^-$ antiport, (c) Na^+—HCO_3^- cotransport, and (d) H^+ efflux via voltage sensitive H^+ channels. These membrane transport processes which result in pH_e changes may be located in both neuronal and glial cell membranes.
3. The post-stimulation alkaline undershoot reflects coupled clearance of K^+ and H^+ by active transport processes involving Na^+/K^+ ATPase.

6.4 Role of Glial Cells in pH_e Homeostasis

Electrical stimulation of the afferent input or depolarization of neurones has been found to evoke intracellular acid shifts (Chesler 1985). Recent recordings from rat cortical astrocytes, however, indicate that electrical stimulation of the cortical surface produces a rapid intracellular alkaline shift which is correlated with the time course of membrane depolarization (Fig. 74; Chesler and Kraig 1989). The pH of astrocytes also becomes more alkaline if exposed to elevated external $[K^+]$ (Chesler and Kraig 1987). During SD in rat neocortex, when the membrane potential is close to 0 mV, the glial membrane potential increases by as much as 0.8 of a pH unit (Fig. 75). The mechanism of the alkaline shift in glial cells is not yet clear, but the basic transport processes which regulate the pH in glia have recently been described (Deitmer and Schlue 1987; Kettenmann and Schlue 1988; Astion and Orkand 1988; Astion et al. 1989). These involve amiloride-sensitive Na^+/H^+ exchange, SITS-sensitive Cl^-/HCO_3^- exchange

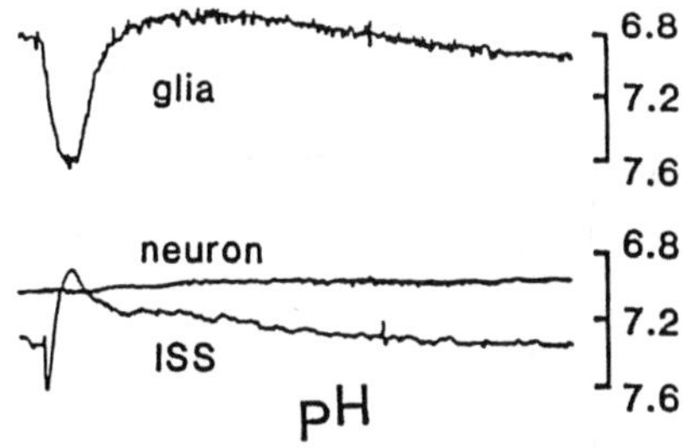

Fig. 75. Changes in pH_e during SD. Records show typical changes (from different experiments) recorded with pH-ISMs from glia, neurones, and interstitial space (*ISS*) in rat neocortex. (From Kraig and Petito 1989)

(astrocytes and neuropile glial cells), and SITS-insensitive Na^+-HCO_3^- cotransport (oligodendrocytes).

In the presence of extracellular Ba^{2+} the acid shifts in the isolated rat spinal cord are diminished (Syková and Jendelová unpublished results) and glial cells are hyperpolarized (Ballanyi et al. 1987; Astion and Orkand 1988; Astion et al. 1989). Chesler and Kraig (1989) found that the stimulation-evoked alkaline shifts in glial cells could not be elicited in the presence of extracellular Ba^{2+}; on the contrary they found slight acidification, despite the fact that $[K^+]_e$ increased in the normal manner. This demonstrated that alkaline shifts are elicited by glial cells depolarization per se. Inwardly directed electrogenic Na^+—HCO_3^- cotransport has been suggested as the mechanism for the depolarization-induced alkaline shift in glial cells of the mud-puppy optic nerve and the leech glia (Astion and Orkand 1988; Astion et al. 1989; Deitmer and Schlue 1989). However, the same transport mechanism in mammalian glia has not yet been confirmed.

It is reasonable to assume that mammalian glial cells regulate the pH_e at the expense of their pH_i. In the spinal cord, as in the optic nerve (Ransom et al. 1985b), the occurrence of stimulation-evoked acid shifts is related to gliogenesis (see Sect. 6.2.2). It was found that, in the spinal cord, the acid shifts were effectively blocked by the CA inhibitor acetazolamide; on the other hand, the alkaline shift, but not the acid shift, was directly related to synaptic activity (Fig. 76). It was found that active snail neurones became acid and that their surface pH_e was at first alkaline and then acid (Thomas 1988a,c). In hippocampal slices, electrical stimulation with a single electric pulse induced a long-lasting extracellular alkaline shift (Krishtal et al. 1987). The initial alkaline shift which is masked by acidification in adult animals was unmasked in the isolated frog spinal cord during the application of acetazolamide (see Figs. 70, 71; Chvátal et al. 1988; Syková 1989). An alkaline shift was also found in the isolated frog spinal cord after the application of excitatory amino acids (Fig. 77; Endres et al. 1986b). The alkaline changes described in the rat spinal cord (Syková and Svoboda 1990), the cerebellum (Kraig et al. 1983; Chesler and Chan 1988), the hippocampus (Krishtal et al. 1987), and the vagus

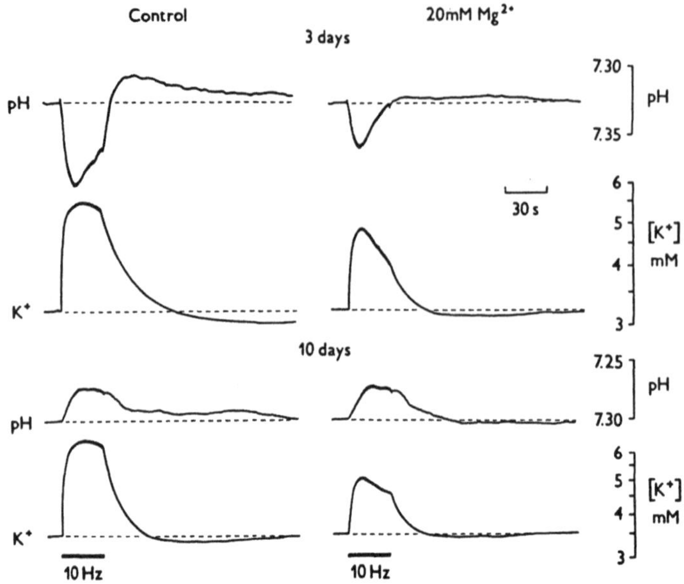

Fig. 76. Stimulation-evoked pH_e and $[K^+]_e$ changes in the spinal dorsal horn of 3- and 10-day-old rats, and effect of 20 mM Mg^{2+} on changes. Note decrease of the alkaline shift, decrease of the post-stimulation acid shifts, and decrease of $[K^+]_e$ changes. Only the acid shift which occurred during stimulation in the 10-day-old rat remained unaffected. (From Jendelová and Syková 1991)

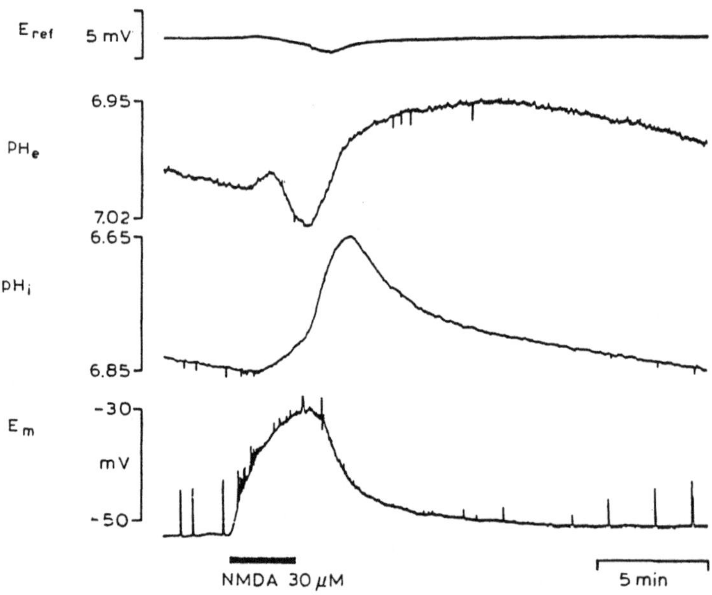

Fig. 77. Changes in pH_e and pH_i around and in a motoneurone of an isolated frog spinal cord during application of NMDA into the bathing solution. In this experiment two double-barrelled pH-ISMs were used. (From Endres et al. 1986b)

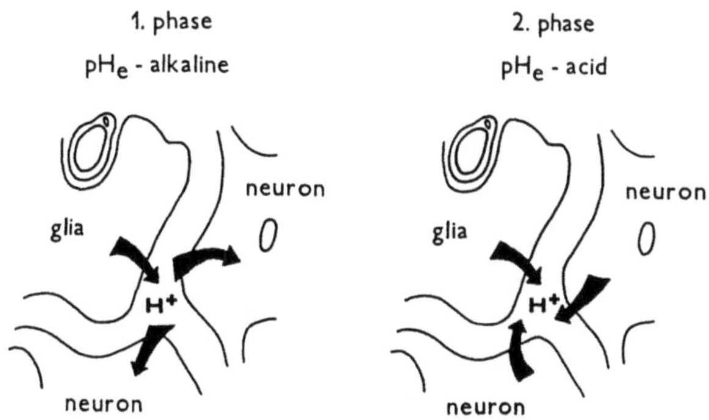

Fig. 78. Hypothetical net flux of H^+ equivalents among neurones, glial cells, and ECS during neuronal activity. 1. phase, at onset of neuronal activity, a net influx of acid equivalents into neurones dominates over an efflux of acid equivalents from glial cells, resulting in an extracellular alkaline shift. 2. phase, with prolonged neuronal activity, a net efflux from both neurones and glia results in an extracellular acid shift. (Adapted from Chesler 1990)

nerve (Endres at al. 1986) were not affected by membrane transport inhibitors or by metabolic blockers, with the exception of Ca^{2+} and H^+ channel blockers. Therefore, there is good reason to suppose that the initial alkaline shifts in adult animals and the alkaline shifts in immature nervous tissue occur in the neuronal microenvironment as a result of net H^+ influx during neuronal and fibre activity. The neurones must eventually extrude the acid by one of the mechanisms proposed in this chapter. The glial cells, however, buffer the alkaline shifts and are at least partly responsible for the long-term post-stimulation acid shifts in the ECS (see Fig. 78; Chesler 1990).

6.5 pH_e Changes in the Retina

Recent studies have shown that stimulation of the retina by light evokes changes in pH_e which are presumably due to the release of acid from retinal cells. In the retina of the drone (Coles et al. 1988) and the frog (Borgula et al. 1987, 1989), alkaline shifts have been evoked by light stimulation. Maintained illumination has evoked both acidification in the proximal retina and alkalinization in the distal retina of the toad (Fig. 79; Oakley and Wen 1989). The light-evoked acid shifts were maximal in the most proximal retina, where there was only very little alkaline shift or undershoot presumably due to the large distance from the source of the light-evoked decrease in pH_e. These findings confirm the early discovery of light-evoked acidification of the inner retina in the pig by Tsacopoulos and Levy (1976). Oakley and Wen (1989) found that a block of synaptic transmission to second order neurones reduced the acid shifts but had literally no effect on alkaline shifts. It is therefore suggested that alkaline shifts are of receptoral origin, while acid shifts are of post-receptoral origin.

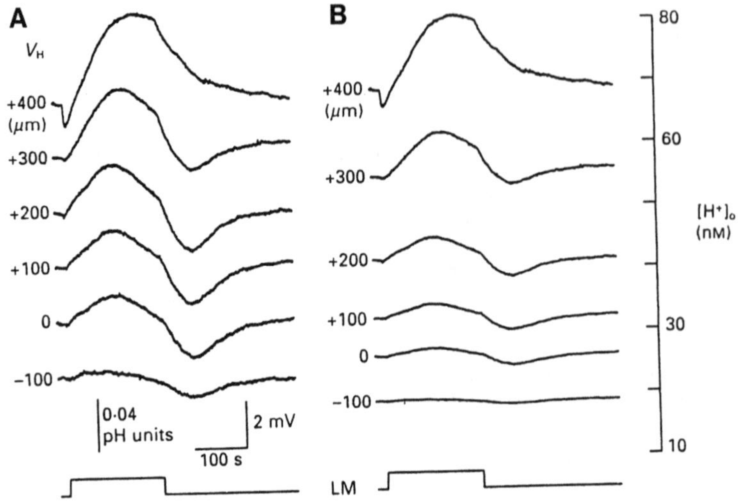

Fig. 79A,B. Light-evoked changes in pH_e and $[H^+]_e$ as a function of retinal depth in the isolated retina of the toad. **A** The retina was superfused with phosphate-buffered solution for more than 1 h. The tip of the pH-ISM was first positioned about 2000 μm below the receptor surface (+400 μm electrode advancement) and the change in electric voltage (V_H) was measured in response to a 180 s light stimulus. The light stimulus is indicated by the upward deflection of the light monitor waveform (*LM*). The electrode was then withdrawn in 100 μm steps, and the same light stimulus was presented at each electrode position (shown next to each response). The baseline level of V_H also was measured at each position, in order to calculate the baseline pH_e at each depth. **B** Light-evoked changes in V_H from **A** were linearized, resulting in the light-evoked changes in $[H^+]_e$. In **B**, the waveforms are placed at the appropriate vertical position to indicate the actual values of $[H^+]_e$. (From Oakley and Wen 1989)

Oakley and Wen (1989) further suggest that retinal neurones extrude H^+ via Na^+/H^+ exchange and Cl^-/HCO_3^- exchange, since both amiloride and SITS caused sustained alkalinization in darkness and decreased light-evoked pH_e changes. Light-evoked changes in pH_e increased after acetazolamide inhibition of CA (Borgula et al. 1989; Oakley and Wen 1989). These findings support the hypothesis that CA, which is localized in the Müller cells (glia) (Musser and Rosen 1973; Linser and Moscona 1984) and to a lesser extent in the horizontal cells, increases the effectiveness of the bicarbonate buffer system. Müller cells therefore not only play an important role in K^+ homeostasis, but also regulate the pH_e in the extracellular fluid surrounding the retinal neurones. Furthermore, in the isolated toad retina, switching from bicarbonate-buffered solution to phosphate-buffered solution (i.e., to solution with a buffering capacity 75 times lower) caused an acid shift in the pH_e and a decrease in light-evoked pH_e changes (Oakley and Wen 1989). Switching to phosphate-buffer also attenuates the b wave of the ERG, which is a measure of the electrical activity of second order retinal neurones (Hagins et al. 1970; Winkler 1986); this is probably due to acidification of the ECS (Oakley and Wen 1989).

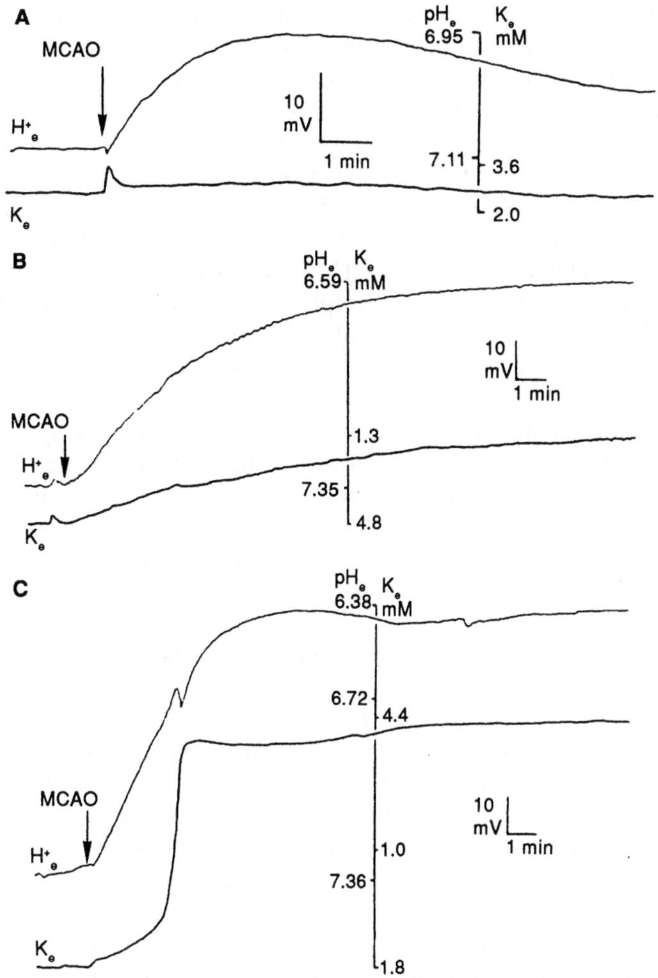

Fig. 80A–C. Three response to middle cerebral artery occlusion in baboon. **A** $[K^+]_e$ marginally increased; significant pH$_e$ shift. **B** Significant pH$_e$ and $[K^+]_e$ shift following middle cerebral occlusion. **C** Major ion shift produced by middle cerebral artery occlusion; massive changes in pH$_e$ and $[K^+]_e$. (From Harris et al. 1987)

6.6 pH$_e$ Changes During Anoxia, Ischaemia, Epilepsy, and SD

In pathological states, disturbances of acid-base balance in the ECS can attain extreme values and cause irreversible changes such as ischaemic necrosis. It has long been known that brain tissue acidosis can occur during complete ischaemia and circulatory arrest (Crowell and Kaufmann 1961; Heuser et al. 1975; Silver 1978; Nemoto and Frinak 1981; Siemkowics and Hansen 1981; Kraig et al. 1983, 1986; Harris and Symon 1984; Harris et al. 1987). Harris et al. (1987) described typical pH$_e$ and $[K^+]_e$ changes in monkey cerebral cortex during occlusion of the middle cerebral artery (Fig. 80). Changes in the pH$_e$

correspond to the degree of ischaemia. During complete ischaemia the changes have a triphasic course. First of all, the pH_e falls by about 0.3 of a pH unit and K^+ almost indiscernibly rises; this is followed by a faster drop in the pH_e, by about 0.6 of a pH unit, and a rise of several millimoles per litre in the $[K^+]_e$. When the K^+ level rises above 10 mmol/l, a break occurs in the increase in pH_e and an alkaline decrease of about 0.1 of a pH unit appears for a few seconds. A sharp increase in K^+ to about 40 mmol/l is accompanied by a further acid shift of several tenths of a pH unit (Harris and Symon 1984; Kraig et al. 1986; Harris et al. 1987). However, when the $[K^+]_e$ increment attains about 10 mmol/l, $[Ca^{2+}]_e$ falls abruptly to about 0.1 mmol/l (Harris and Symon 1984; Kraig et al. 1986). The pH_e change in rat frontal cerebral cortex during complete ischaemia is directly proportional to the simultaneously recorded tissue pCO_2, dc potential, and decrease in blood pressure (Kraig et al. 1986). In the rat spinal cord, discontinuation of artificial respiration caused asphyxia, a decrease in blood pressure, and pH_e changes which had a similar triphasic course to that described, accompanied by a two-step increase in $[K^+]_e$ (Syková and Svoboda 1990).

The contemporary conception of the mechanism of pH_e changes in pathological states is based primarily on findings which show that lactic acid – which contributes to acidosis – accumulates in the ECS. Eklof and Siesjö (1972) showed that the lactate concentration in the rat brain rose after arterial occlusion lasting only 5 min. At a blood pressure of 70 mmHg, it amounted to 20 mmol/kg wet weight. This indicates that there is sufficient lactate to cause the acid changes observed during the first few minutes' ischaemia. This conception is reinforced by experiments which show that the decrease in pH_e during ischaemia is a progressive process which is dependent on the brain blood flow and the blood glucose level (Siemkowicz and Hansen 1981).

During partial ischaemia, the pH_e and blood pressure are soon stabilized, without any marked changes in $[K^+]_e$; this testifies to the fact that the pH_e is the resultant of dynamic equilibrium between the formation (release) of H^+ and its buffering, which is assured chiefly by HCO_3^- (Kraig et al. 1985b, 1986), or by clearance by active transport. An exponential increase in pH_e may be due to progressive loss of the buffering capacity of the ECS, increased H^+ production, or reduced clearance of CO_2 and lactic acid from the blood. However, it is also possible (although experimental evidence is lacking) that since the intracellular space is about 0.1–0.2 of a pH unit more acid than the ECS, acidosis may be due to an increase in membrane permeability for H^+. The mechanism of the alkaline shift observed on the exponential curve of the decrease in pH_e is still unclear. Kraig et al. (1983) suggested that the alkaline shifts might be caused by a decrease in $[Ca^{2+}]_e$.

In the final phase of ischaemic acidosis induced by lactic acid, tissue necrosis develops and the cell membranes seem to lose their permeability for H^+ and other ions, which leads to further acid changes. Kraig et al. (1986) proposed two factors which might contribute to the loss of membrane permeability for H^+: (a) the ischaemia-damaged transmembrane Na^+ and Cl^- gradient and the loss of brain HCO_3^- reserves – which impairs regulation of pH_e by way of

Na^+/H^+ exchange and Cl^-/HCO_3^- exchange, and (b) a decrease in the diffusion permeability of the plasma membranes for H^+, due to the binding of a raised amount of H^+ to the surface proteins.

6.7 Functional Significance of pH_e Changes

Studies carried out over the past few years have shown that pH_e changes which either accompany neuronal activity or are observed in pathological states can influence impulse transmission in the CNS. The original conception that changes in the pH_e influence only enzymatic reactions has now been supplemented by many further possibilities regarding the functional significance of dynamic pH_e changes.

Changes in the pH_e have a pronouced effect on ionic channels. Large pH_e changes suppress the ionic currents through Na^+ channels (Hille 1968); currents running through Ca^{2+} channels are, however, suppressed by much smaller $[H^+]_e$ increments (Iljima et al. 1986). Rapid pH_e changes over a range of a few tenths of a pH unit activate Na^+ currents, which may temporarily depolarize sensory neurones (Krishtal and Pidoplichko 1980, 1981a,b; Konnerth et al. 1987). Konnert et al. (1987) showed that this proton-induced current is carried primarily by Na^+ through Ca^{2+} channels. An increase in the pH_e probably transforms Ca^{2+} channels from a voltage-gated state to a proton-gated state. This alters the permeability of the Ca^{2+} channel, so that Na^+ transport has priority. Krishtal and Pidoplichko (1980, 1981a,b) suggested that the proton-activated increase in membrane permeability for Na^+, as the response to a drop in the pH_e, could act as a signal of nociceptive stimulation. These changes have already been demonstrated in the sensory neurones of spinal ganglia and the same type of receptor is probably localized in nerve endings, where it can act as a pH sensor. About half of the sensory neurones in the trigeminal ganglia were found to be proton sensitive; the majority were small neurones with a diameter of less than 26 μm. Proton receptors exhibited partial or complete desensitization. These findings indicate that proton receptors play an important role in the transmission of nociceptive information.

When applied iontophoretically to isolated mammalian spinal neurones, H^+ raised the threshold for action potential formation and suppressed responses to application of the neurotransmitters glutamate, glycine, and GABA (Gruol et al. 1980). However, $[H^+]$ changes had a dual effect on neurones in tissue cultures; the excitability of neurones both increased and decreased, probably owing to inhibition of their sensitivity to various amino acids. The pH_e also influences the charge of amino acid molecules, their affinity to receptors, and the permeability of ionic channels.

Balestrino and Somjen (1987, 1988) showed that pH_e changes influenced the excitability of neurones in the hippocampus. Orthodromically evoked focal action potentials (population discharges) were suppressed during hypercapnia, but stimulated during hypocapnia. These results indicate that H^+ in the extracellular fluid has an inhibitory effect on the excitability of neurones in the hippocampus.

Reduction of the pH_e to 6.6 pH units inhibited Na^+/K^+ ATPase activity in astrocyte tissue cultures (Walz and Hinks 1987) and in the membrane fraction of the rat spinal cord (Hájek et al. 1988). At the same time, K^+ uptake by the cells was reduced by 60%. These results show that pH_e changes influence K^+ homeostasis in the CNS.

Changes in pH_i or pH_e are also closely related to the regulation of cell volume (see Sect. 7.4). In experiments on the isolated frog spinal cord, an acid change of 0.4 of a pH unit in the Ringer solution significantly reduced the size of the ECS, while alkaline changes increased its size (Chvátal and Syková, unpublished results).

In pathological states like ischaemia, anoxia, and nervous tissue injury, acidosis is considered to be an important factor in the functional recovery of nervous tissue (Rehncrona et al. 1980; Siesjö 1981, 1984). Astrup et al. (1977) suggested that there was an association between a drop in pH_e and the loss of somatosensory evoked potentials in primates. Acidosis is also a probable cause of irreversible nerve cell injury (Kraig et al. 1987). It is interesting that glial cells in ischaemic brain become twice as acid as the interstitial fluid and thus evidently act as buffer elements which safeguard pH_e homeostasis. It can thus be concluded that the maintenance of pH_i and pH_e homeostasis in the CNS is fundamental to the normal functioning of the CNS. In addition, oscillations in the pH_e act as an important modulating and controlling factor, influencing the excitability of cells, as well as their metabolism and proliferative capacity. The full functional significance of dynamic pH_e changes is still an open question.

7 Dynamic Changes in Size of the ECS

The function and survival of the brain depends on the regulation of nervous tissue volume. Brain volume can change when the size of one of the three brain compartments – intracellular, extracellular or, intravascular – changes. These changes can result from a whole series of pathological events. For example, enlargement of the intracellular and extracellular space owing to the accumulation of water leads to brain oedema (Pappius 1974). If water accumulates intracellularly, it causes cytotoxic oedema (Klatzo 1967). Impairment of the blood-brain barrier and the accumulation of extracellular fluid rich in proteins leads to vasogenic oedema. Dynamic, reversible changes in the size of glial cells (and probably neurones and nerve fibres as well) also occur during neuronal activity, even under physiological conditions. Enlargement of cells is compensated by a decrease in the size of the ECS, without any increase in the total amount of water in the nervous tissue. These processes alter the size of the brain compartments simply as a result of a water shift from one compartment to another, so that brain oedema does not develop.

Using the optic method, Lipton (1973) observed reversible enlargement of the cells in cerebral cortical slices 1 s after applying a train of excitatory impulses. In the frog brain, Van Harreveld and Trubatch (1974) and Trubatch et al. (1977) used measurements made in the electron microscope to show that, after depolarization with KCl solution, water shifted from the extracellular to the intracellular space; the size of postsynaptic structures increased from 4.8% to 14.2% of tissue volume. Using a highly sensitive piezoelectric sensor, Iwasa et al. (1980) and Tasaki and Iwasa (1982) recorded an increase in the size of crab nerve axones and squid giant axones during action potentials, while Tasaki and Burne (1983) found swelling of the cells in the dorsal ganglia and enlargement of neuronal elements in the spinal cord of the frog during sciatic nerve stimulation. These studies show that volume changes occur in the elements and ECS of the nervous system during excitation, but they do not provide any information on the size or dynamics of these changes.

The method of determining the volume of the ECS using K^+-ISMs allowed dynamic measurement of changes during electrical and adequate stimulation.

7.1 Measurement of Changes in Size of the ECS by Means of K^+-ISMs

ISMs allow the monitoring of ion concentrations in real time and at a specific tissue site, and thereby act as indicators of ion diffusion in tissue. In the CNS, diffusion in the ECS is influenced by its size (α) and by its tortuosity (λ) – factors described in Sect. 3.3.3. Using ISMs, Phillips and Nicholson (1979)

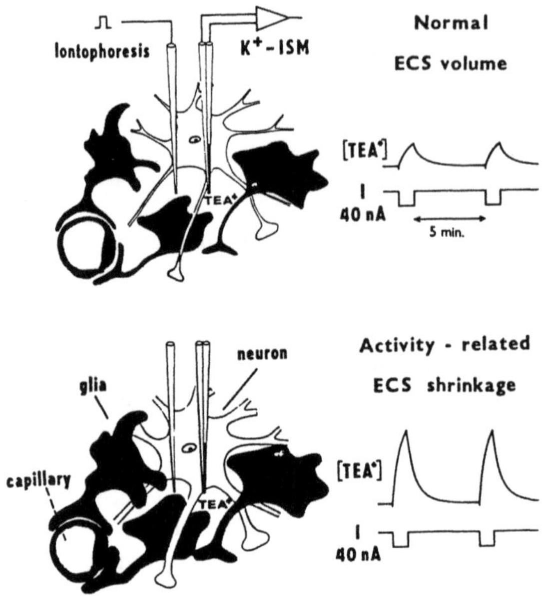

Fig. 81. Experimental set-up when measuring dynamic ionic and volume changes in the ECS

demonstrated changes in the ECS during SD. During neuronal activity after electrical stimulation, ECS changes were found in the cat sensorimotor cortex (Dietzel et al. 1980, 1982), the rat optic nerve (Ransom et al. 1985a,b), the isolated frog spinal cord (Syková 1987; Syková et al. 1987a), and the rat spinal cord in vivo (Svoboda et al. 1989, 1990; Svoboda and Syková 1989, 1991). A decrease in the size of the ECS was also observed in the rat spinal cord during chronic nociceptive stimulation (see Sect. 7.3).

Dynamic changes in the size of the ECS are studied by means of iontophoretic administration of ions which do not cross the cell membranes and therefore remain in the ECS. Their concentration in the ECS is in inverse proportion to the size of the space (Nicholson et al. 1979). The ion exchanger for K^+ is highly sensitive to tetraalkylammonium ions and choline, which, in small concentrations, are not toxic and do not cross cell membranes until after at least 60 s contact (Phillips and Nicholson 1978; Kříž and Syková 1981). Choline and tetraethylammonium (TEA^+) or tetramethylammonium (TMA^+) ions are therefore used for testing changes in the size of the ECS. Figure 81 represents the principle employed in the measurement of relative dynamic changes in the size of the ECS in the CNS. Using a current of 10–100 nA, the ions are administered for 5–60 s into the ECS with a single or multichannel iontophoretic pipette. The tip of the K^+-ISM which records ion concentration changes during administration is 40–200 μm distant from the tip of the iontophoretic pipette. If this distance needs to be known exactly – for instance, to determine changes in λ or α of ECS – the two electrodes must be glued together with dental

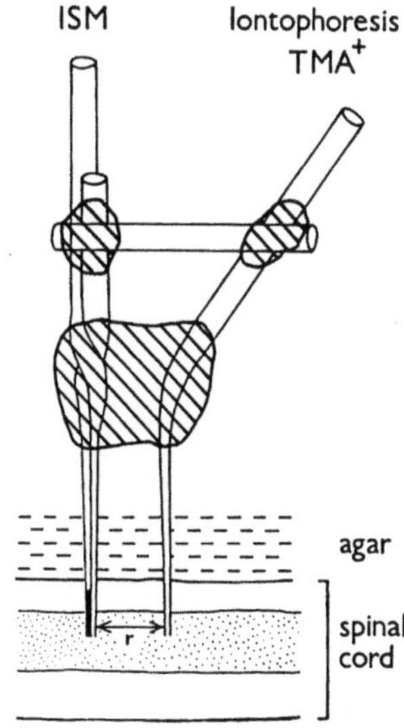

Fig. 82. Experimental arrangement for determining ECS volume fraction (α) and tortuosity (λ) in the spinal cord in vivo. Two microelectrodes, the double-barrelled microelectrode sensitive to K^+ and to TMA^+ (*ISM*) and the micropipette for TMA^+ iontophoresis, were glued together with dental cement at their shanks to stabilize the intertip distance (r), and connected at their upper end with a glass tube bridge

cement and the inter-tip distance measured exactly under the microscope. After they have been joined together, the electrodes are introduced into the nervous tissue. The ions are administered at regular intervals (1–10 min), always using the same current. The few values obtained before starting stimulation or the relevant experiment are regarded as the control if they do not differ from one another by more than 5%. Relative changes in the size of the ECS can be computed from the change in concentration of the given ions, according to the formula:

$$\text{Percentage decrease in ECS} = \left(1 - \frac{\Delta[\text{TEA}^+]\ \text{prior to activity}}{\Delta[\text{TEA}^+]\ \text{after activity}}\right) \times 100$$

The same method can be used for computing and determining changes in λ and α, although, for determination, it is best to administer the ions into the nervous tissue for 60–80 s and to know the exact inter-tip distance of the two electrodes. When measurement in the nervous tissue has been completed, the microelectrodes are withdrawn into a thick layer of 0.3% agar which has been poured over the surface of the nervous tissue (see Fig. 82); the procedure is then repeated. Measurement of the concentration change in the agar gives us the free diffusion values for the relevant ion, in which $\lambda = 1$ and $\alpha = 1$. K^+-ISMs are calibrated in appropriate calibration solutions containing TEA^+, TMA^-, or choline. The diffusion coefficient values, λ and α, in the brain or spinal cord are determined by computation after Nicholson and Phillips

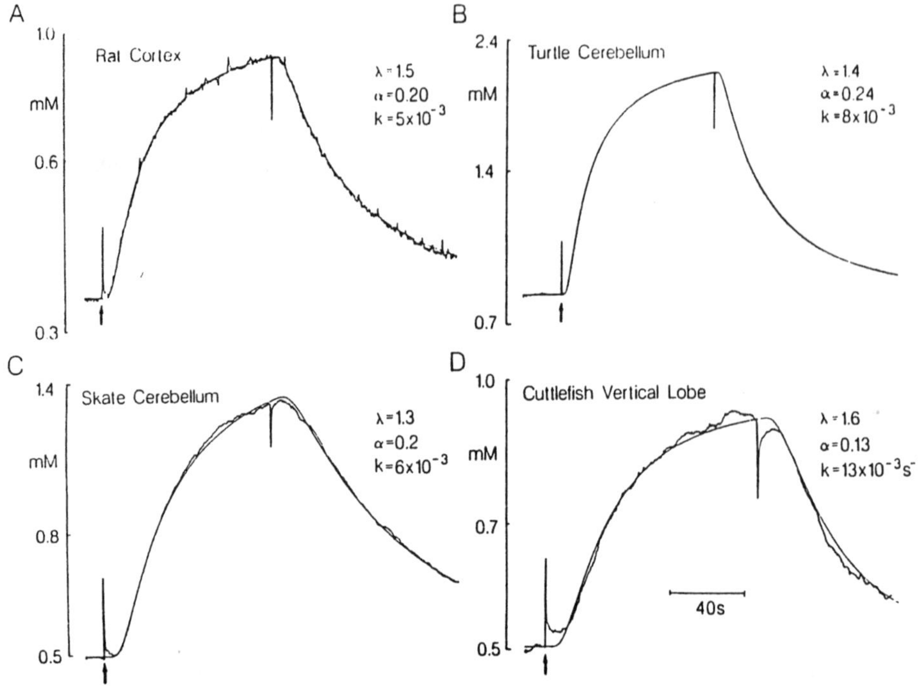

Fig. 83A–D. Diffusion profiles for TMA^+ in four species using iontophoresis. In each case iontophoresis commenced at the time indicated by *arrow*. Duration of iontophoresis was 90 s (**A**), 80 s (**B**), 90 s (**C**), and 112 s (**D**). The time scale (see **D**) is the same for each species but the amplitude calibrations, which are logarithmic, differ. In each case the theoretical curve was matched to the experimental data by varying α, λ, and the uptake (k) until the best fit was obtained, the values of these parameters that optimized the curve fit are shown in *inset*. Measurements in **A**, **C**, and **D** were recorded in vivo in anaesthetized animals; those in **B** were recorded in an isolated cerebellum. In all experiments a continuous forward bias current was applied to the iontophoresis electrode to maintain tip concentration and provide a stable baseline. (From Nicholson and Rice 1990)

(Phillips and Nicholson 1979; Nicholson and Phillips 1981; see Sect. 3.3.3). As stated in Section 3.3.3, the resting λ and α values in rat dorsal horns do not differ significantly from those measured in the cerebellum (Nicholson and Phillips 1981; Nicholson and Rice 1988). ECS volume in rat dorsal horns amounts to 0.24 ± 0.01 (23% ± 5%) and λ amounts to 1.54 ± 0.04 (mean ±SD of mean; $n = 21$). The resting λ and α values at different depths in the dorsal horns did not differ significantly from one another.

The parameters α and λ are geometric in nature. Nicholson and Rice (1990) showed that a third factor, uptake (k), should be introduced into the diffusion equation to represent non-specific loss of ions across the cell membranes due to the concentration gradient, leakiness, or some other, as yet unknown, form of uptake (see also Nicholson and Phillips 1981). Figure 83 summarizes α, λ and k in rat cortex, turtle cerebellum, skate cerebellum, and cuttlefish vertical lobe.

From the diffusion profiles, it is apparent that the factors affecting diffusion of ions in the ECS are not significantly different either between species or in different areas of the brain and spinal cord.

7.2 Changes Induced in Size of the ECS by Electrical Stimulation

Enlargement of cell volume has been demonstrated in the cerebral and cerebellar cortex during SD (Van Harreveld and Khattab 1967; Hansen and Olsen 1980; Phillips and Nicholson 1979), and during perfusion of the cerebral cortex with a solution with a high K^+ concentration (Bourke and Nelson 1972; Lipton 1973). During epileptic activity in a chronic gliotic focus, K^+ homeostasis is impaired and the size of the ECS diminishes in the neighbourhood of the focus. At the site of maximum accumulation of K^+ during paroxysmal activity (5–7 mmol/l), the ECS volume was reduced by 35.5 ± 16.5%, whereas in areas of gliotic tissue, where $[K^+]_e$ rises by only about 2 mmol/l, ECS volume rose by 25.2% ± 18.3% (Heinemann et al. 1983; Heinemann and Dietzel 1984; Dietzel and Heinemann 1986).

Changes in cell size accompanied by diminution of the ECS have, however, been demonstrated under physiological conditions without massive depolarization of the tissue, huge ion changes, block of transmission, paroxysmal activity, or SD. Using K^+-ISMs, local ECS volume changes and their dynamics can be studied during adequate and electrical stimulation. Repetitive direct stimulation of the cat sensorimotor cortex, or stimulation of the ventroposterolateral thalamic nucleus causes ECS volume to be reduced by up to 50% (Dietzel et al. 1980, 1982). Similar ECS volume changes were observed in the isolated frog spinal cord (Syková 1987; Syková et al. 1987a) and in the rat spinal cord in vivo (Svoboda et al. 1989; Svoboda and Syková 1989, 1991) during repetitive electrical stimulation of the dorsal root or a peripheral nerve.

Electrical stimulation of dorsal roots at an intensity supramaximal for the activation of low-threshold fibres, with a frequency of 1–100 Hz, and lasting 30 s–3 min, caused a 10%–50% reduction in the size of the ECS in frog dorsal horns. The higher the stimulation frequency and intensity, and the greater the concomitant changes in $[K^+]_e$, the greater was the decrease in ECS size (Fig. 84A and B). The decrease persisted after stimulation was discontinued and often lasted longer than the stimulation-evoked increase in $[K^+]_e$. Figure 84 shows that the decrease in the size of the ECS after 60 s stimulation of the dorsal roots with a frequency of 100 Hz persisted 30 min after stimulation. If the same preparation was stimulated again (180 s) after recovery, the decrease in the size of the ECS lasted for up to 90 min (Syková 1987). The longer the stimulation time, the longer the duration of dynamic changes in ECS volume, even though there were no further marked changes in $[K^+]_e$. Comparison of the course of $[K^+]_e$ changes with reduction of the ECS shows that the time course of the latter is much longer than that of the former. It is also apparent that volume changes develop more slowly than changes in $[K^+]_e$ and often do not attain the maximum change until $[K^+]_e$ changes have disappeared (Fig. 84A,B).

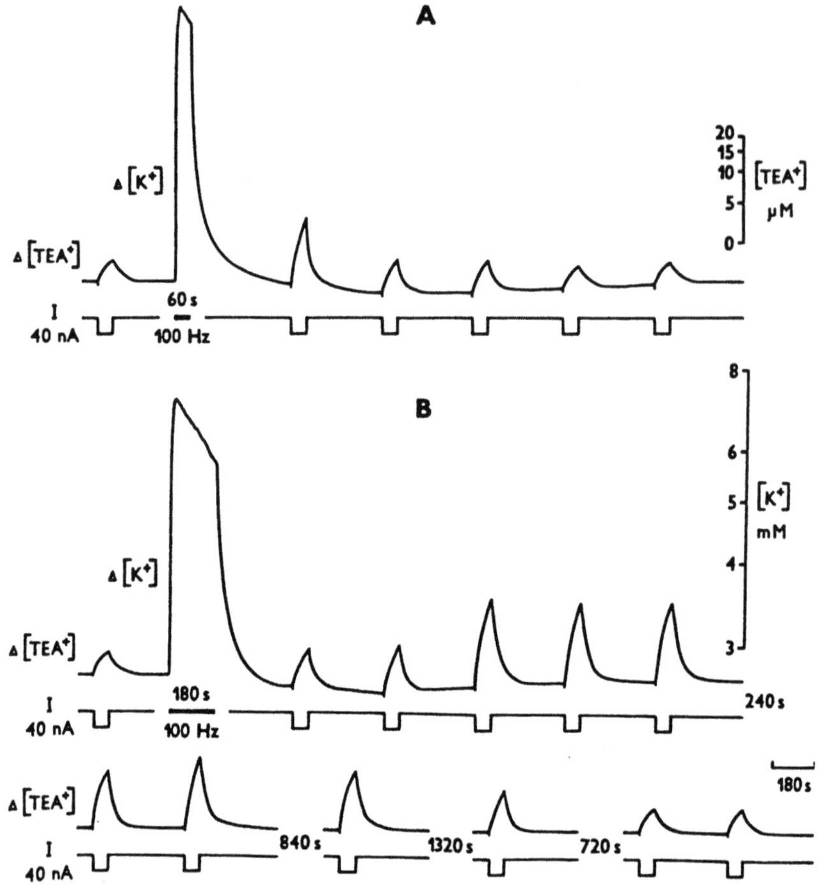

Fig. 84A,B. Effect of repetitive electrical stimulation of dorsal root VIII on the size of the ECS in segment VIII of the isolated frog spinal cord. Recordings from a depth of 300 μm. **A** 60 s stimulation (frequency 100 Hz). **B** 3 min stimulation at the same frequency. The iontophoretic current (40 nA, bottom recordings) was applied with a micropipette filled with 5 mmol/l tetraethylammonium chloride (TEA^+) solution. The top recordings show changes in [TEA^+] during iotophoretic administration and changes in $[K^+]_o$ induced by stimulation. The numbers 240 s, 840 s, 1320 s and 720 s between the traces indicate the time intervals in seconds, which are not shown in the figure. Recordings **A** and **B** are from the same experiment. (From Syková 1987)

Diminution of the volume of the ECS was also found in the rat lumbar cord during electrical stimulation of a hind limb nerve. After repetitive electrical stimulation, by means of two needles introduced below the skin of the paw, we observed up to a 30% reduction in ECS volume in the dorsal horns of segment L_4 (Figs. 85 and 88). The actual decrease depended on the intensity, frequency, and length of stimulation, but in contrast to the changes observed in the isolated frog spinal cord, they did not last for more than 15–20 min.

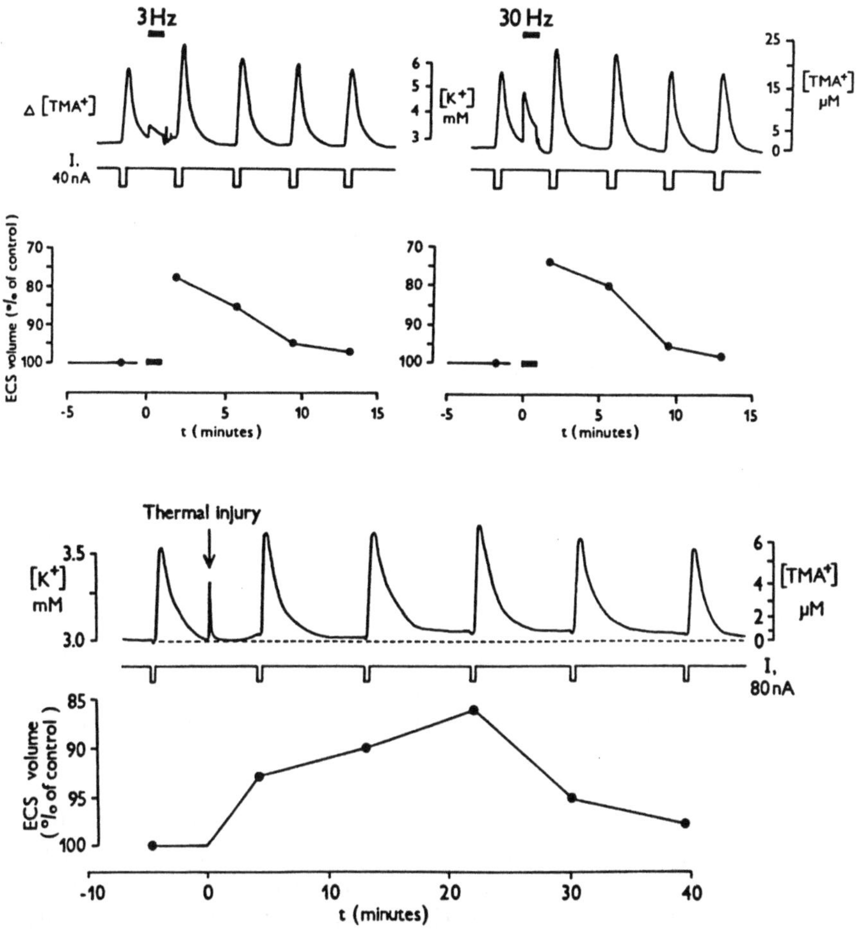

Fig. 85. Shrinkage of ECS induced in the rat spinal dorsal horn by tetanic electrical stimulation of the ipsilateral hind paw (frequency 3 or 30 Hz) and after burning a paw by applying hot oil (thermal injury). The *top* recordings show changes in the tetramethylammonium chloride ($\Delta[TMA^+]$) concentration during iontophoretic administration with a current of 40 nA (I) and $[K^+]_e$ changes during stimulation. The graphs illustrate the decrease in the ECS after stimulation as a percentage of the control. The control (100%) is the mean pre-stimulation value for three curves differing from one another by not more than 5%. (From Svoboda and Syková 1991)

The decrease in ECS volume has a characteristic depth profile in both the sensorimotor cortex of the cat (Dietzel et al. 1980, 1982) and the spinal cord of the frog and rat (Svoboda and Syková 1989, 1991). Maximum changes were always observed at sites of greatest neuronal activity, i.e., where maximum $[K^+]_e$, $[Ca^{2+}]_e$, and pH_e changes were found. Figure 86 shows the distribution of changes in ECS volume in the isolated frog spinal cord induced by 60 s stimulation of dorsal root VIII with a frequency of 10–30 Hz. Maximum changes were observed at a depth of about 300–350 μm, i.e., in the region

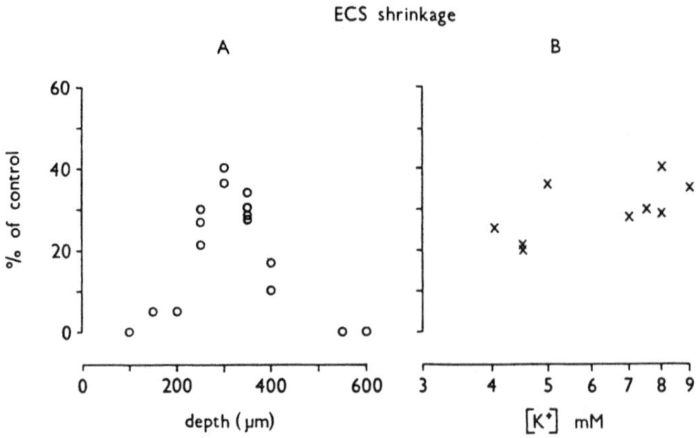

Fig. 86. A Shrinkage of ECS induced by dorsal root stimulation (30 Hz, 60 s) at different depths of the dorsal horns of the isolated frog spinal cord. **B** Graphic illustration of the decrease in the ECS volume in correlation to the change in $[K^+]_e$ evoked by stimulation; recordings at a depth of 250–350 μm. Every *circle* or *cross* is the result of a single experiment. (Syková and Chvátal, an unpublished figure)

where ionic changes and the number of interneurones per tissue volume unit were greatest. At this depth the size of the ECS was reduced by at least 20%, even though the change in $[K^+]_e$ was only 1 mmol/l (Fig. 86). It is evident from our experiments that ionic changes are not the only factor which influences dynamic changes in the ECS.

Our experiments also demonstrated that reduction of ECS volume in the spinal cord during repetitive stimulation also occurred during a block of synaptic transmission (Syková and Chvátal, unpublished results). When synaptic activity was blocked, dynamic changes in ECS volume were 50%–70% smaller. We can thus assume that the decrease in the size of the ECS is due partly to activation of primary afferent fibres. During stimulation, the fibres or neighbouring Schwann cells (or both these elements) swell. Ransom et al. (1985a,b) demonstrated dynamic volume changes in the rat optic nerve (Fig. 87). They observed that a decrease in the size of the ECS could not be seen during development until myelination started. A long-term increase in the size of the dorsal ganglia, the axones, and their endings in the spinal cord was also observed during prolonged depolarization of primary afferent fibres by Tasaki and Iwasa (1982) and Tasaki and Byrne (1983), who used a piezoelectric sensor to detect mechanical changes.

The ouabain block of Na^+/K^+ pump activity markedly prolongs and augments changes in $[K^+]_e$ and pH_e (Figs. 5, 66, 67), even when synaptic activity is blocked by the divalent cations Mg^{2+} and Mn^{2+}. In the isolated frog spinal cord, ouabain had a pronounced effect on dynamic volume changes produced in the ECS by repetitive electrical stimulation; it doubled the decrease in the size of the ECS and lengthened its duration (Syková and Chvátal, unpublished results). These findings show that changes in metabolic activity, as well as ionic

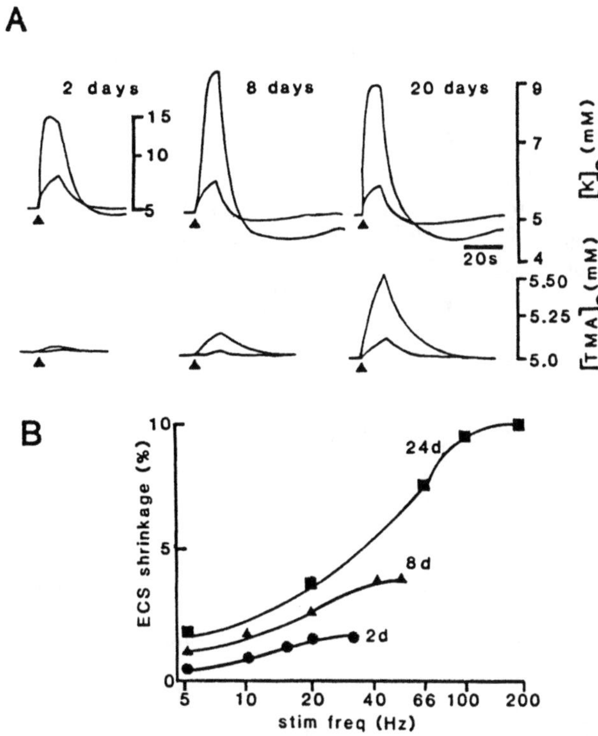

Fig. 87A,B. Activity-dependent shrinkage of ECS in rat optic nerve during development. **A** Age dependency of evoked increases in $[K^+]_e$ (normal Ringer's solution, *top traces*) or $[TMA^+]_e$ (5 mM TMA added to Ringer's solution, *bottom traces*). Responses to 10 s trains at two frequencies are shown for nerves at three different ages; 2 days (5 and 20 Hz), 8 days (5 and 40 Hz), and 20 days (10 and 100 Hz). Maximal evoked increase in $[TMA^+]_e$ progressively increased with age. Note the separate $[K^+]_e$ scale for 2-day-old nerve. **B** Graph showing maximum ECS shrinkage (percentage) as a function of stimulus frequency (10 s trains) for nerves of three different ages. The maximum frequency which the optic nerve can follow increases with age. (From Ransom et al. 1985b)

changes, play an important role in the mechanism of dynamic volume changes in the cellular elements and the ECS during activation processes in the CNS.

7.3 Changes Induced in Size of the ECS by Adequate Stimulation

In the bee's retina, repeated flashes of light cause the ECS to shrink by up to 32% (Orkand et al. 1984). This finding indicates that synaptic clefts narrow during physiological photoreceptor activity. In the dorsal horns of the rat spinal cord, nociceptive stimuli and injury produce long-term ionic changes, especially in $[K^+]_e$ and $[H^+]_e$ (see Sects. 4.1.4 and 6.2.2.1); these changes are the result of persistent hyperactivity of the interneurones in the dorsal horns

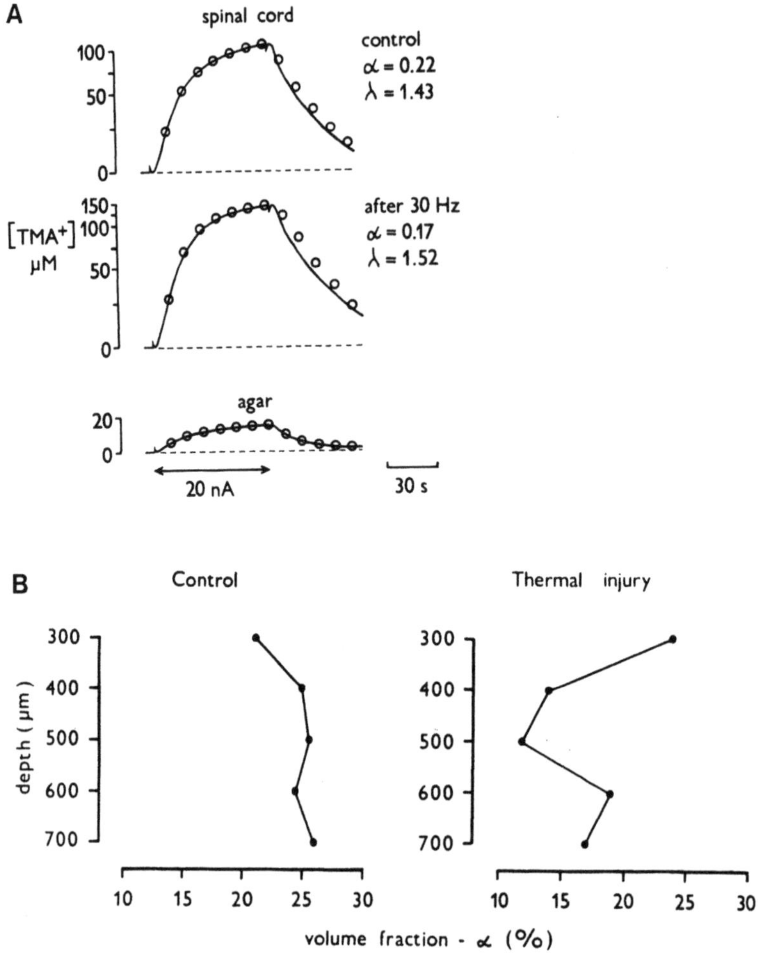

Fig. 88. **A** Changes in the TMA^+ concentration in segment L_4 of the rat spinal cord at a depth of 600 μm (*spinal cord*) and after withdrawing the paired electrodes into agar poured over the surface of the cord (*agar*). TMA^+ was applied iotophoretically for 60 s at regular intervals (10 min) with a 20 nA current. Changes in $[TMA^+]$ were recorded with the K^+-ISM. Both electrodes were fixed firmly together, with their tips 220 μm apart. The diffusion curve recorded 9 min after stimulation is larger, due to the ECS volume decrease. The volume fraction and tortuosity before stimulation (*control*) a_c = 0.22 and a λ = 1.68.; and after stimulation (30 Hz) a_a = 0.17 and λ = 1.75. The open symbols are the theoretical values calculated for this experiment. **B** Effect of thermal injury on the ECS volume at various depths in dorsal horn. *Left*, typical ECS volume fraction at various depths of the spinal dorsal horn in the non-stimulated animal before injury. *Right*, the ECS volume decreased at a depth of 400–700 μm. The maximal ECS volume decrease from about 24% to about 12% was found at depth 500 μm 10–25 min after the thermal injury. (After Svoboda and Syková 1991)

(Svoboda et al. 1988; Syková and Svoboda 1990) and it is therefore not surprising that this activity is also associated with changes in the size of the ECS.

The application of a few drops of boiling water or hot oil to the skin of the rat's hind paw was followed by injury which was attended by a typical local inflammatory reaction. Figures 85 and 88 show that this injury was accompanied by a decrease in the size of the ECS which began immediately after injury and attained the maximum 15–25 min after application of the nociceptive stimulus. The decrease in the ECS volume amounted to up to 50% and persisted for 40–60 min after injury. Weaker nociceptive stimuli, such as a drop of hot water or squeezing a few toe joints with forceps, induced distinct diminution of the ECS volume, but not for more than 5 min. The decrease in the size of the ECS after nociceptive stimuli and injury had a typical depth profile; it was greatest in sites of maximum increase in neuronal activity and maximum changes in $[K^+]_e$, $[H^+]_e$, and $[Ca^{2+}]_e$.

The above studies demonstrate that there are important dynamic changes in cell and ECS volume during physiological stimuli. The question remains whether changes in the size of the ECS are a physiologically significant factor in the interaction between cells, in ionic homeostasis, or in synaptic transmission.

7.4 Mechanisms of Dynamic Changes in Size of the ECS

Enlargement of cell volume constitutes a serious complication of a great many forms of acute and chronic injury and impairment of the function of the CNS. Brain oedema occurs not only during hypoxia, ischaemia, and injury, but also during osmotic imbalance, intoxication, and metabolic disturbances (Baethman 1978, 1979; Kempski 1986). An increase in cell size can also result from neuronal activity, evoked by physiological stimuli, as described in the preceding chapter. This chapter deals with those mechanisms which lead to enlargement of the intracellular compartment of the brain, accompanied by compensatory reduction of the size of the ECS. It does not deal with disturbances of the intravascular compartment or with that mechanism of volume changes which involves an increase in total water volume in the nervous tissue. The mechanisms so far suggested, and their participation in the volume changes observed, merely contribute basic information; their actual role in the development of volume changes still requires further research. The mechanisms of volume changes in pathological states are probably very different to those which come into play during stimulation-evoked neuronal activity.

The modern concept of the maintenance of cell volume was summed up by MacKnight and Leaf (1977) as follows: Under steady state conditions, intra- and extracellular dissolved solids and liquids are in double Donnan's equilibrium, in which the osmotic pressure of intracellular non-permeable anions (proteins), attracting positively charged ions, is balanced and compensated for by the active extrusion of sodium and the low Na^+ permeability of the cell membranes. The three most important factors are: (1) the number of negative charges on intracellular macromolecules, (2) selective membrane permeability for biologically important ions, and (3) the active sodium transport system – the Na^+/K^+ pump activated by Na^+ and by K^+-activated, Mg^{2+}-dependent ATPase.

An effective feedback system which helps control cell volume evidently exists. Control mechanisms are set in motion by "signals" announcing a change in cell volume, e.g., $[Na^+]_i$ (Robinson 1968), cellular pH (Cala 1983), and $[Ca^{2+}]_i$ (Cala 1983). We know that the activity of the ion pump and membrane permeability alter if cell volume changes. Kempski (1986) described four main mechanisms of enlargement of cell volume and reduction of the ECS:

1. Enlargement of cell volume is the result of a disturbance of Donnan's equilibrium owing to an increase in membrane permeability for Na^+. This gives rise to a transcellular osmotic gradient.
2. Inhibition of systems regulating cell volume, e.g., the Na^+/K^+ pump or other transport systems.
3. A disturbance of the meaning of signals announcing a change in cell volume, e.g., if the signal is not identified.
4. An increase in the volume of glial cells and neurones in states which the organism can temporarily tolerate, e.g., in the premitotic phase and during neuronal activity.

The mechanism of cell enlargement and reduction of the ECS during activation of neuronal elements probably includes even more mechanisms.

7.4.1 Volume Changes Induced by Changes in Extracellular Osmolarity

A decrease in extracellular osmolarity is always associated with the entry of H_2O into the cells. Since animal cells are highly permeable for H_2O, osmotic relationships between the extra- and intracellular environment are very quickly equalized. Nevertheless, some cells – including glial cells and neurones – are able to activate, within a few seconds, regulatory mechanisms capable of reducing any enlargement to the original value, even in a hypotonic environment (Chaussy et al. 1981; Kempski et al. 1983). Volume regulation is associated with an increase in oxygen uptake by glial cells – evidence that regulatory mechanisms are energy dependent (Kempski et al. 1983) – and is accompanied by an increase in permeability for K^+. Cala (1983) suggested that cation flow is due to exchange for H^+, but Cl^- shifts are due to exchange for HCO_3^-. It can therefore be assumed that volume regulation processes are attended by changes in the intra- and extracellular pH. Ca^{2+} also probably plays an important role in the regulation of volume changes, via Ca^{2+}-activated K^+ channels (Cala 1983; Eckert and Tillotson 1978).

In hypertonic solutions, cell volume diminishes. We do not yet know whether nervous tissue cells are capable of active volume regulation in hypertonic solutions. The mechanism appears similar to the one in evidence when there is volume increase in solutions with a raised K^+ concentration (see Sect. 7.4.3).

7.4.2 Volume Changes During Neuronal Activity

Lipton (1973) showed that the size of the elements in cerebral cortex slices increased during the first second after the application of a volley of excitatory impulses. Enlargement was reversible and was accompanied by an increase in tissue respiration. Volume enlargement was blocked by exchanging Cl^- for impermeable anions.

The size of volume changes after repetitive stimulation (Dietzel et al. 1980, 1982; Ransom et al. 1985a,b; Syková 1987; Svoboda et al. 1989, 1991) can be deduced from changes in tissue osmolarity and from ion shifts across cell membranes. It is evident that, in nervous tissue, the volume of neurones, nerve fibres, and glial cells increases at the expense of the ECS. We still do not know the proportion of the volume increase in the individual elements either during neuronal activity or in pathological states. We anticipate that the molecular mechanisms of volume changes in neuronal elements are different from those in glial cells.

Dietzel et al. (1982) investigated the mechanism of reduction of ECS volume which accompanies changes in the ionic composition of the microenvironment of nerve cells in cat cerebral cortex. Their results indicate that intraneuronal osmolarity can rise by up to 30 mosmol/l because, owing to splitting of large molecules during increased metabolism of neurones, the number of intracellular particles rises; this leads to the uptake of H_2O and to an increase in cell volume. They estimated that a 30 mosmol/l increase in intraneuronal osmolarity would reduce the ECS by less than 10%, and also suggested that the decrease in the size of the ECS depended on a second mechanism – enlargement of glial cell volume owing to the K^+ shifts mediated by the spatial buffer mechanism. The glial cells become larger in places where K^+ accumulates and smaller in places where the K^+ concentration is low. Experiments on the isolated frog spinal cord, which showed that the size of the ECS diminished only in areas of maximum neuronal activity and maximum K^+ accumulation, whereas in neighbouring areas it increases, support this hypothesis (Syková 1987; Syková and Chvátal, unpublished results). The question remains as to whether, under physiological conditions, the uptake of NaCl by the neurones participates in volume changes, as suggested by Kraig and Nicholson (1978) during SD. Ion shifts leading to an increase in glial cell volume do not comprise only the spatial buffer mechanism alone, however, and are therefore discussed separately.

7.4.3 Transport Systems of Glial Cells and Regulation of Their Volume

Bourke et al. (1976, 1978) were the first to show that the incubation of brain slices in medium with a high K^+ concentration (50 mmol/l), in the presence of bicarbonates, led to enlargement of astroglia volume. In a study published in 1969, Bourke reported active K^+ uptake, which was later confirmed in the glial cells by other authors (Hertz 1978; Waltz and Hertz 1982). It was found that

K^+ levels of over 3 mmol/l stimulated glial metabolism (Orkand et al. 1973; Somjen et al. 1976). The ATP concentration in glial cells, but not in neurones, fell by up to 50% during an increase in $[K^+]_e$ (Schousboe et al. 1970). Studies over the past 15 years have shown that the volume increase is not the result of a K^+ shift alone, and that complex transport processes are involved in volume regulation – particularly SITS-sensitive anion exchange, amiloride-sensitive Na^+/H^+ exchange, and furosemide-sensitive Na^+-K^+-Cl^- cotransport (Johnson et al. 1982; Kimelberg 1979, 1981; Kimelberg et al. 1979a,b; Kimelberg and Frangakis 1985; Waltz and Hertz 1984; Wolpaw and Martin 1984; Chassande et al. 1988). Ransom et al. (1985a,b) demonstrated a decrease in the size of the ECS in the optic nerves of adult rats during stimulation, although no such decrease was seen in the nerves of newborn rats, which were not yet myelinated (see also Fig. 87). The decrease was correlated to the presence of Cl^- and was blocked by SITS and furosemide. It is clear from existing studies that changes in glial cell volume under physiological and pathophysiological conditions are undoubtedly the result of shifts of H_2O, cations, and anions across glial membranes (a) passively (K^+, Cl^-), (b) by a shift requiring enzymatic energy (Na^+, Cl^-, K^+ in the presence of HCO_3^-), and (c) by the K^+ spatial buffer mechanism.

It is also possible that enlargement of glial cells is due to the massive release of excitatory transmitters. These stimulate the release of K^+, the accumulation of which leads to further transmitter release, giving rise to a positive feedback process leading – especially under pathological conditions – to massive transmitter accumulation (Hertz 1977). Glutamate causes depolarization of postsynaptic endings and enlarges their volume (Van Harreveld 1966), and the volume of nearby glial cells (Hösli and Hösli 1976; Waniewski and Martin 1983). The uptake of 1 mmol/l glutamate is associated with the uptake of 2 mmol/l Na^+. The redistribution of K^+ and Na^+ resulting from the release of glutamate requires energy and if the situation is complicated further (e.g., by anoxia or ischaemia), the regulation of cell volume may be impaired (Kempski et al. 1982). Enlargement of glial cell volume can also lead to the release of norepinephrine, adenosine, and histamine as a result of increased entry of Cl^- into the cells and activation of CA (Bourke et al. 1983).

7.4.4 Changes in Cell Volume Induced by Inhibition of Na^+/K^+ ATPase

The injection of ouabain into brain tissue showed that a block of Na^+/K^+ ATPase leads to a drastic increase in cell volume (Cornog et al. 1967). Impairment of Na^+ and K^+ transport in neurones and glial cells can be an important factor in the origin and regulation of dynamic volume changes. In our experiments on the isolated frog spinal cord, the administration of ouabain led to a decrease in ECS volume after stimulation which was twice that observed when Na^+/K^+ ATPase activity was not blocked (Syková and Chvátal, unpublished results); the dynamic change in the ECS also lasted longer. If we compare the duration of dynamic volume changes in the isolated frog spinal cord with their duration in the rat spinal cord in vivo, we see that, in frog preparations – in which the experiments were conducted at a lower temperature (16–20°C) – the decrease in the size of the ECS after stimulation lasted three times longer than it did in the rat in vivo. The activity of the Na^+/K^+ pump is, therefore, probably an important factor in the mechanism of the regulation of cell volume.

7.5 Functional Significance of Dynamic Volume Changes in the Microenvironment of Nerve Cells

It is clear from the results presented in this chapter that significant changes occur in ECS volume, not only in pathological states, but also under physiological conditions. It must, however be said that we do not know the functional significance of these changes and can only present hypotheses which have been proposed or emerge from the findings. Van Harreveld and colleagues showed that the dendritic spines in the frog brain increased in size, and they grew shorter but thicker (Trubatch et al. 1977). Since the impedance of the elements is directly proportional to their length and in inverse proportion to their diameter, the impedance of dendritic spines decreases as the effectiveness of synaptic transmission increases (Rall 1974). Changes observed in the shape and dimensions of postsynaptic elements may, therefore, provide a morphological basis for the facilitation of transmission in nervous tissue.

In experiments on preparations consisting of the isolated frog spinal cord connected to an isolated hind limb, we observed that lowering of the threshold for the flexor reflex after repetitive tetanic transcutaneous stimulation of the hind limb followed the same time course as the decrease in the ECS (Syková 1987). These measurements indicate that post-stimulation diminution of the ECS is a cause of hyperexcitability and synaptic hyperactivity in the dorsal spinal horns.

According to Van Harreveld and Fifková (1975), fluid uptake by the dendritic spines is the basis of long-term potentiation of transmission occurring after electrical stimulation. An increase in synaptic transmission activity may also be one of the mechanisms leading to plastic neuronal changes (Rall 1974). Lastly, the given volume changes may influence both the formation of short-term memory traces (Van Harreveld and Fifková 1974; Patel and Stewart 1988) and long-term memory, e.g., through their effect on proteosynthesis (Van Harreveld and Fifková 1974; Cherkin et al. 1976).

8 Conclusion

The cellular elements fill the greater part of the brain and spinal cord and are separated by only narrow intercellular spaces, which are rich in ions, proteins, transmitters, and enzymes. The intercellular spaces are known as the ECS or microenvironment, the existence of which has been neglected by science for many years. Studies in the 1960s first gave rise to the conception of the ECS as a fluid with a complex structure and composition, in a space exactly circumscribed by cell membranes (Schmitt and Samson 1969), and is only in the last 10 years that the ECS has come to be recognized as a dynamically changing microenvironment for nerve cells, with a significant influence on the function of the nervous system.

The interaction between the intracellular compartment, the extracellular compartment, CSF, and the vascular bed unquestionably plays an important role in the normal functioning of the CNS. I have described many of the changes in composition and size of the ECS (the most important are illustrated in Fig. 89) and have indicated their possible functional significance. Ion fluxes across the cell membranes and the release of transmitters and neuromodulators obviously continually alter the composition of the ECS. In addition to increasing our knowledge of the function of glial cells and their influence on the composition and size of the extracellular space, we have also established a firmer basis on which to define the properties, function, and significance of the microenvironment of nerve cells. This new concept of the microenvironment and new findings on its relationship to the other compartments of the CNS make a substantial contribution towards a better understanding of the development of pathological states and the required treatment.

Two subjects which require further research are the physiological significance of dynamic changes in the microenvironment and the mechanism of the maintenance of ionic homeostasis in the intercellular spaces in the CNS. The latter must be assured continuously in order that signals may be formed in, and transmitted from neurones. Disturbances of the control of ionic homeostasis very quickly lead to nervous tissue damage and are therefore dreaded complications of anoxia, ischaemia, and brain oedema. We should not forget that the microenvironment also guarantees the intercellular transmission of metabolic substrates, enzymes, transmitters, and hormones. Oxygen and glucose are also transported to the cellular elements, and CO_2 is removed from them, via the microenvironment. Lastly, the fundamental question of the extent to which the microenvironment assures intercellular interaction and integration is still unanswered. Does it provide a kind of "wireless" connection alongside the now familiar and well-described connection by nerve fibres and synapses? To help resolve this question we must distinguish the ways in which substances

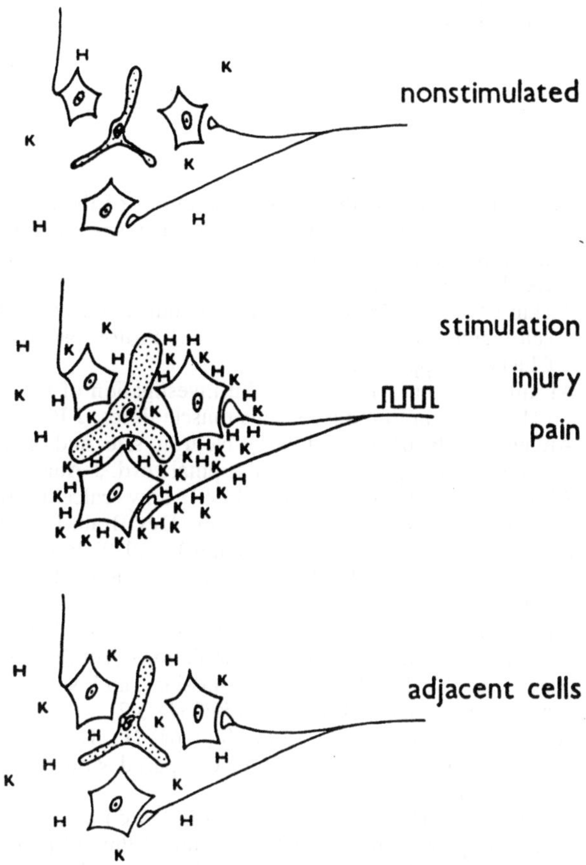

Fig. 89. Dynamic changes in ECS volume and $[K^+]_e$ and $[H^+]_e$ changes in unstimulated CNS tissue, in the immediate vicinity of stimulated cells, and in the neighbourhood of stimulated cells (*adjacent cells*)

released into the ECS move within it in order to reach their destination.

The studies cited in this monograph demonstrate that in the nervous system – as in any other community – the viability and function of the individual (the neurones and glial cells) is dependent on the microenvironment, which, in turn, is strongly influenced by the activity of the individual (the nerve cells). Only a harmonious interrelationship provides optimum conditions for the function of the nervous system.

9 References

Adelman WJ, Fitzhugh R (1975) Solutions of the Hodgkin-Huxley equations modified for potassium accumulation in periaxonal spaces. Fed Proc 34:1322–1329

Adey WR (1969) Slow electrical phenomenon in the central nervous system. Neurosci Res Program Bull 7:75–180

Aickin CC (1988) Movement of acid equivalents across the mammalian smooth muscle cell membrane. In: Thomas RC (ed) Proton passage across cell membranes. Wiley, Chichester, pp 3–12

Aickin C, Thomas RC (1977) An investigation of the ionic mechanism of intracellular pH regulation in mouse soleus muscle fibres. J Physiol (Lond) 273:295–316

Aitken JT, Bridger JE (1961) Neuron size and neuron population density in the lumbosacral region of the cat's spinal cord. J Anat 95:38–53

Albrecht D, Heinemann U (1989) Low calcium-induced epileptiform activity in hippocampal slices from infant rats. Dev Brain Res 48:316–320

Albrecht D, Rausche G, Heinemann U (1989) Reflections of low calcium epileptiform activity from area CA1 into dentate gyrus in the rat hippocampal slice. Brain Res 480:393–396

Aleksandrovskaya MM, Chizhenova RA (1979) Quantitative analysis of glial elements in the sensorimotor cortex on stimulation of cerebral structures. Fiziol Zh 65:1207–1213

Alger BE, Teyler TJ (1978) Potassium and short-term response plasticity in the hippocampal slice. Brain Res 159:239–242

Ames A III, Sakanone M, Endo S (1964) Na, K, Mg, and Cl concentrations in choroid plexus fluid and cisternal fluid compared with plasma ultrafiltrate. J Neurophysiol 27:672–681

Ames A III, Higashi K, Nesbett FB (1965) Relation of potassium concentration in choroid plexus fluid to that in plasma. J Physiol (Lond) 181:506–515

Ammann D (1986) Ion selective microelectrodes. Principles, design and application. Springer, Berlin Heidelberg New York

Ammann D, Lanter F, Steiner RA, Schulthess P, Shijo Y, Simon W (1981) Neutral carrier based hydrogen ion selective microelectrode for extra- and intracellular studies. Anal Chem 53:2267–2269

Ammann D, Oesch U, Bührer T, Simon W (1987) Design of ionophores for ion-selective microsensors. Can J Physiol Pharmacol 65:879–884

Anderson SA (1975) Pain control by sensory stimulation. In: Liebskind JC, Bonica JJ, Albe-Fessard DG (eds) Advances in pain research and therapy, vol 3. Raven, New York, pp 569–585

Armington JC (1974) The electroretinogram. Academic, New York

Astion ML, Orkand RK (1988) Electrogenic Na^+/HCO_3^- cotransport in neuroglia. Glia 1:355–357

Astion ML, Obaid AL, Orkand RK (1989) Effects of barium and bicarbonate on glial cells of *Necturus* optic nerve. J Gen Physiol 93:731–744

Astrup J, Norberg K (1976) Potassium activity in cerebral cortex in rats during progressive severe hypoglycemia. Brain Res 103:418–423

Astrup J, Symon L, Branston NM, Lassen NA (1977) Cortical evoked potential and extracellular K^+ and H^+ at critical levels of brain ischemia. Stroke 8:51–57

Astrup J, Rehncrona S, Siesjö BK (1980) The increase in extracellular potassium concentration in the ischemic brain in relation to the preischemic functional activity and cerebral metabolic rate. Brain Res 199:161–174

Baethmann A (1978) Pathophysiological and pathochemical aspects of cerebral edema. Neurosurg Rev 1:85–100

Baethmann A (1979) The mechanical, circulatory, osmotic and toxic genesis of brain edema. In: Ahnefeld FW, Bergmann H, Burri C (eds) Der bewusstlose Patient. Springer, Berlin Heidelberg New York, pp 56–75

Balestrino M, Somjen GG (1987) Effects of CO_2 and interstitial pH changes on synaptic transmission in CA1 region of rat hippocampus in vitro. In: Schurr A, Teyler TJ, Tseng MT (eds) Brain slices: fundamentals, applications and implications. Karger, Basel, pp 147–151

Balestrino M, Somjen GG (1988) Concentration of carbon dioxide, interstitial pH and synaptic transmission in hippocampal formation of the rat. J Physiol (Lond) 396:247–266

Ballanyi K, Grafe P, Ten Bruggencate G (1987) Ion activities and potassium uptake mechanisms of glial cells in guinea-pig olfactory cortex slices. J Physiol (Lond) 382:159–174

Banna NR, Saade NF, Atweh SF, Jabbur SJ (1986) Prolonged discharge of wide-dynamic-range spinal neurons evoked by formaldehyde injected in their cutaneous receptive fields. Exp Neurol 93:275–278

Barker JL, Nicoll RA, Padjen A (1975) Studies on convulsants in the isolated frog spinal cord. II. Effects on root potentials. J Physiol (Lond) 245:537–548

Barres BA, Chun LLY, Correy DP (1988) Ion channel expression by white matter glia I. Glia 1:10–30

Barron DH, Matthews BHC (1935) Intermittent conduction in the spinal cord. J Physiol (Lond) 85:73–103

Barron DH, Matthews BHC (1938) The interpretation of potential changes in the spinal cord. J Physiol (Lond) 92:276–321

Baumgarten CM (1981) An improved liquid ion exchanger for chloride ion-selective microelectrodes. Am J Physiol 241:C258–C263

Baylor DA, Nicholls JG (1969) Changes in extracellular potassium concentration produced by neuronal activity in the central nervous system of the leech. J Physiol (Lond) 203:555–569

Ben-Ari Y, Krnjevič K, Reinhardt W (1979) Hippocampal seizures and failure of inhibition. Can J Physiol Pharmacol 57:1462–1466

Benninger C, Kadis J, Prince DA (1980) Extracellular calcium and potassium changes in hippocampal slices. Brain Res 187:165–182

Benos DJ (1982) Amiloride: a molecular probe of sodium transport in tissues and cells. Am J Physiol 242:C141–C145

Bernard C (1878) Leçons sur les phénoménes de la vie. Baillieve, Paris

Besson J-M, Chaouch A (1987) Peripheral and spinal mechanisms of nociception. Physiol Rev 67:67–186

Bevan S, Chiu SY, Gray PTA, Ritchie JM (1985) The presence of voltage-gated sodium, potassium and chloride channels in rat cultured astrocytes. Proc R Soc (Lond) B225:299–313

Blank WF Jr, Kirshner HS (1977) The kinetics of extracellular potassium changes during hypoxia and anoxia in the cat cerebral cortex. Brain Res 123:113–124

Borgula GA, Karwoski CJ, Steinberg RH (1987) Light-evoked pH changes in the retina of the *in vitro* frog eyecup. Invest Ophthalmol Vis Sci [Suppl] 406

Borgula GA, Karwoski CJ, Steinberg RH (1989) Light-evoked changes in extracellular pH in frog retina. Vision Res 29(9):1069–1077

Bormann J, Hamill OP, Sakmann B (1987) Mechanism of anion permeation through channels gated by glycine and γ-aminobutyric acid in mouse cultured spinal neurons. J Physiol (Lond) 385:243–286

Boron WF, Boulpaep EL (1983) Intracellular pH regulation in the renal proximal tubule of the salamander. Na-H exchange. J Gen Physiol 81:29–52

Boron WF, Siebens AW, Nakhoul NL (1988) Role of monocarboxylate transport in the regulation of intracellular pH of renal proximal tubule cells. In: Thomas RC (ed) Proton passage across cell membranes. Wiley, Chichester, pp 91–105
Bourke RS (1969) Evidence for mediated transport of chloride in cat cerebral cortex in vitro. Exp Brain Res 8:219–231
Bourke RS, Nelson KM (1972) Further studies on the K^+-dependent swelling of primate cerebral cortex in vivo: the enzymatic basis of the K^+-dependent transport of chloride. J Neurochem 19:663–685
Bourke RS, Kimelberg HK, Nelson LR (1976) The effects of temperature and inhibitors of HCO_3^- stimulated swelling and ion uptake of monkey cerebral cortex. Brain Res 105:309–323
Bourke RS, Kimelberg HK, Dazé MA (1978) Effects of inhibitors and adenosine on (HCO_3^-/CO_2)-stimulated swelling and Cl^--uptake in brain slices and cultured astrocytes. Brain Res 154:196
Bourke RS, Kimelberg HK, Dazé M, Church G (1983) Swelling and ion uptake in cat cerebrocortical slices: control by neurotransmitters and ion transport mechanisms. Neurochem Res 8:5–24
Bradbury MWB, Davson H (1965) The transport of potassium between blood, cerebrospinal fluid and brain. J Physiol (Lond) 181:151–174
Bradbury MWB, Kleeman CR (1967) Stability of the potassium content of cerebrospinal fluid and brain. Am J Physiol 213:519–528
Bradbury MWB, Štulcová B (1970) Efflux mechanism contributing to the stability of the potassium concentration in cerebrospinal fluid. J Physiol (Lond) 208:415–430
Branston NM, Strong AJ, Symon L (1977) Extracellular potassium activity, evoked potential and tissue blood flow. J Neurol Sci 32:305–321
Brinley FJ, Kandel ER, Marshall WH (1960a) Effect of gamma-aminobutyric acid (GABA) on K^+ outflow from rabbit cortex. J Neurophysiol 23:237–245
Brinley FJ, Kandel ER, Marshall WH (1960b) Potassium outflow from rabbit cortex during spreading depression. J Neurophysiol 23:246–256
Brown AG, Kirk RJ, Martin HF (1973) Descending and segmental inhibition of transmission through the spinocervical tract. J Physiol (Lond) 230:689–705
Brownell WE, Bader CR, Bertrand D, de Ribaupierre Y (1985) Evoked mechanical responses of isolated cochlear outer hair cells. Science 227:194–196
Bruggencate ten GT, Lux HD, Liebl L (1974) Possible relationship between extracellular potassium activity and presynaptic inhibition in the spinal cord of the cat. Pflugers Arch 349:301–317
Bruggencate ten GT, Nicholson C, Stöckle H (1976) Climbing fiber evoked potassium release in cat cerebellum. Pflugers Arch 367:107–109
Bührle CP, Sonnhof U (1981) Ion fluxes across the membrane of motoneurones during the action of glutamate. In: Syková E, Hník P, Vyklický L (eds) Ion-Selective Microelectrodes and their use in excitable tissues. Plenum, New York, pp 187–190
Bureš J, Burešová O (1981) Cerebral $[K^+]_e$ increase as an index of the differential susceptibility of brain structures to terminal anoxia and electroconvulsive shock. J Neurobiol 12:211–220
Bureš J, Burešová O, Křivánek J (1974) The mechanism and applications of Leão's spreading depression of electroencephalographic activity. Academic, New York
Cala PM (1983) Volume regulation by red blood cells: Mechanisms of ion transport. Mol Physiol 4:33–52
Cameron IR, Caronna J (1976) The effect of local changes in potassium and bicarbonate concentration on hypothalamic blood flow in the rabbit. J Physiol (Lond) 262:415–430
Cammer W, Sacchi R, Sapirstein V (1985) Immunocytochemical localization of carbonic anhydrase in the spinal cords of normal and mutant (Shiverer) adult mice with comparisons among fixation methods. J Histochem Cytochem 33:45–54

Campbell JN (1981) Examination of possible mechanisms by which stimulation of the spinal cord in man relieves pain. Appl Neurophysiol 44:181–186
Campenot RB (1984) Inhibition of nerve fiber regeneration in cultured sympathetic neurons by local high potassium. Brain Res 293:159–163
Carlini WG, Ransom BR (1986) Regional variation in stimulated extracellular pH transients in the mammalian CNS. Soc Neurosci Abstr 12:452
Castellucci VF, Goldring S (1970) Contributions to steady potential shifts of slow depolarization in cells presumed to be glia. Electroencephalog. Clin Neurophysiol 28:109–118
Cervero F, Iggo A, Ogawa H (1976) Nociceptor-driven dorsal horn neurones in the lumbar spinal cord of the cat. Pain 2:5–24
Chassande O, Frelin C, Farahifar D, Jean T, Lazdunski M (1988) The $Na^+/K^+/Cl^-$ cotransport in C6 glioma cells. Properties and role in volume regulation. Eur J Biochem 171:425–433
Chaussy L, Baethmann A, Lubitz W (1981) Electrical sizing of nerve and glia cells in the study of cell volume regulation. In: Cervos-Navarro J, Fritschka E (eds) Cerebral microcirculation and metabolism. Raven, New York, pp 29–40
Cherkin A, Eckardt MJ, Gerbrant XX (1976) Memory: proline induces retrograde amnesia in chicks. Science 193:242–244
Chesler M (1985) PhD dissertation, New York University Graduate School of Arts and Sciences
Chesler M (1986) Regulation of intracellular pH in reticulospinal neurones of the lamprey, petromyzon marinus. J Physiol (Lond) 341:241–261
Chesler M (1987) pH regulation in the vertebrate central nervous system; microelectrode studies in the brain stem of the lamprey. Can J Physiol Pharmacol 65(5):986–993
Chesler M (1990) The regulation and modulation of pH in the nervous system. Prog Neurobiol 34:401–427
Chesler M, Chan CY (1988) Stimulus-induced extracellular pH transients in the in vitro turtle cerebellum. Neuroscience 27:941–948
Chesler M, Kraig RP (1987) Intracellular pH of astrocytes increase rapidly with cortical stimulation. Am J Physiol 253:R666–R670
Chesler M, Kraig RP (1989) Intracellular pH transients of mammalian astrocytes. Neuroscience 9(6):2011–2019
Chesler M, Nicholson C (1985) Regulation of intracellular pH in vertebrate central neurons. Brain Res 325:313–316
Chmykhova NM (1982) Intracellular potassium activity in frog motoneurons. Sechenov Physiol J USSR LXVIII 3:337–343
Chvátal A, Jendelová P, Kříž N, Syková E (1988) Stimulation-evoked changes in extracellular pH, calcium and potassium activity in the frog spinal cord. Physiol Bohemoslov 37:203–212
Cody AR, Russell IJ (1985) Outer hair cells in the mammalian cochlea and noise-induced hearing loss. Nature 315:662–665
Cohen MW (1970) The contribution by glial cells to surface recordings from the optic nerve of an amphibian. J Physiol (Lond) 210:565–580
Cohen MW, Gerschenfeld HM, Kuffler SW (1968) Ionic environment of neurones and glial cells in the brain of an amphibian. J Physiol (Lond) 197:363–380
Coles JA (1981) Changes in K^+ activity in photoreceptors and glial cells of the drone retina when $[K^+]$ in the superfusate is changed. J Physiol (Lond) 318:16P–17P
Coles JA, Orkand RK (1983) Modification of potassium movement through the retina of the drone (Apis mellifera ♂) by glial uptake. J Physiol (Lond) 340:157–174
Coles JA, Orkand RK (1984) Chloride enters receptors and glia in response to light or raised external potassium in the retina of the bee drone, Apis mellifera. J Physiol (Lond) 357:11P

Coles JA, Tsacopoulos M (1979) Potassium activity in photoreceptors, glial cells and extracellular space in the drone retina. J Physiol (Lond) 290:525–549
Coles JA, Giovannini P, Thomas RC (1988) Changes in extracellular pH induced by light stimulation in slices of honeybee drone retina. J Physiol (Lond) 59P
Collewijn H, Schadé JP (1962) Potassium movements in the cerebral cortex during asphyxia. Acta Morphol Neerl Scand 5:11–20
Conner JD, Detwiler PB, Sarthy PV (1985) Ionic and electrophysiological properties of retinal Müller (glial) cells of the turtle. J Physiol (Lond) 362:79–92
Connors BW, Ransom BR, Kunis DM, Gutnick MJ (1982) Activity-dependent K^+ accumulation in the developing rat optic nerve. Science 216:1341–1343
Cooke JD, Quastel DMJ (1973) The specific effect of potassium on transmitter release by motor nerve terminals. J Physiol (Lond) 228:435–458
Cornog JL, Gonatas NK, Feierman JR (1967) Effects of intracerebral injection of ouabain on the fine structure of rat cerebral cortex. Am J Pathol 51:573–590
Cragg B (1979) Brain extracellular space fixed for electron microscopy. Neurosci Lett 15:301–306
Cragg P, Patterson L, Purves MJ (1977) The pH of brain extracellular fluid in the cat. J Physiol (Lond) 272:137–166
Crawford AC, Fettiplace R (1981) Non-linearities in the responses of turtle hair cells. J Physiol (Lond) 315:317–338
Crowell JD, Kaufmann BN (1961) Changes in tissue pH after circulatory arrest. Am J Physiol 200(4):743–745
Cummins CJ, Glover RA, Selliger OZ (1979) Astroglial uptake is modulated by extracellular K^+. J Neurochem 33:779–785
Cummins J, Hydén H (1962) Adenosine triphosphate levels and adenosine triphosphatases in neurone, glia, and neuronal membranes of the vestibular nucleus. Biochim Biophys Acta 60:271–283
Curtis HJ, Cole KS (1942) Membrane resting and action potentials from the squid giant axon. J Cell Comp Physiol 19:135–144
Czèh G, Syková E, Vyklický L (1980) Neurones activated from nociceptors in the spinal cord of the frog. Neurosci Lett 16:257–262
Czèh G, Kříž N, Syková E (1981) Extracellular potassium accumulation in the frog spinal cord induced by stimulation of the skin and ventrolateral columns. J Physiol (Lond) 320:57–72
Davidoff RA, Hackman JC, Osorio I (1980) Amino acid antagonists do not block the depolarizing effects of potassium ions on frog primary afferents. Neuroscience 5:117–126
Davidoff RA, Hackman JC, Holohean AM, Vega JL, Zhang DX (1988) Primary afferent activity, putative excitatory transmitters and extracellular potassium levels in frog spinal cord. J Physiol (Lond) 397:211
Davis PK, Carlini WG, Ransom BR, Black JA, Waxman SG (1987) Carbonic anhydrase activity develops postnatally in the rat optic nerve. Dev Brain Res 31:291–298
Davson H (1967) Physiology of the cerebrospinal fluid. Churchill, London
Davson H (1976) The blood-brain barrier. J Physiol (Lond) 255:1–28
Davson H, Welch K (1971) The permeation of several materials into the fluids of the rabbit's brain. J Physiol (Lond) 218:337–351
Deitmer JW, Ellis D (1980) Interactions between the regulation of the intracellular pH and sodium activity of sheep cardiac Purkinje fibres. J Physiol (Lond) 304:471–488
Deitmer JW, Schlue WR (1987) The regulation of intracellular pH by identified glial cells and neurones in the central nervous system of the Leech. J Physiol (Lond) 388:261–283
Deitmer JW, Schlue WR (1989) An inwardly directed electrogenic sodium-bicarbonate co-transport in leech glial cells. J Physiol (Lond) 411:179–194

Del Castilo J, Katz B (1954) The membrane change produced by the neuromuscular transmitter. J Physiol (Lond) 125:546–565
Deschenes M, Feltz P (1976) GABA-induced rise of extracellular potassium in rat dorsal root ganglia: an electrophysiological study in vivo. Brain Res 118:494–499
Deschenes M, Feltz P, Lamour Y (1976) A model for an estimate in vitro of the ionic basis of presynaptic inhibition: an intracellular analysis of the GABA-induced depolarization in rat dorsal root ganglia. Brain Res 118:486–493
Dichter MA, Herman CJ, Selzer M (1972) Silent cells during interictal discharges and seizures in hippocampal penicillin foci. Evidence for the role of extracellular K^+ in the transition from the interictal state to seizures. Brain Res 48:173–183
Dick E (1979) Light- and dark-dependent extracellular K^+ activity modulations in the vertebrate retina: Origins of electroretinographic components. PhD thesis, State University of New York, Buffalo
Dick E, Miller RF (1985) Extracellular K^+ activity changes related to electroretinogram components. I. Amphibian (I-type) retinas. J Gen Physiol 85:885–909
Dietzel I, Heinemann U (1986) Dynamic variations of the brain cell microenvironment in relation to neuronal hyperactivity. Ann NY Acad Sci 481:72–84
Dietzel I, Heinemann U, Hofmeier G, Lux HD (1980) Transient changes in the size of the extracellular space in the sensorimotor cortex of cats in relation to stimulus-induced changes in potassium concentration. Exp Brain Res 40:432–439
Dietzel I, Heinemann U, Hofmeier G, Lux HD (1982) Stimulus-induced changes in extracellular Na^+ and Cl^- concentration in relation to changes in the size of the extracellular space. Exp Brain Res 46:73–84
Dowling JE, Ripps H (1976) Potassium and retinal sensitivity. Brain Res 107:617–622
Ebbesson SOE (1976) Morphology of the spinal cord of frog. In: Llinás R, Precht W (eds) Neurobiology. Springer, Berlin Heidelberg New York, pp 679–706
Eccles JC (1964) The physiology of synapses. Springer, Berlin Heidelberg
Eccles JC, Malcolm JM (1946) Dorsal root potentials of the spinal cord. J Neurophysiol 9:139–160
Eccles JC, Eccles RM, Magni F (1961) Central inhibitory action attributable to presynaptic depolarization produced by muscle afferent volleys. J Physiol (Lond) 159:147–166
Eccles JC, Kostyuk PG, Schmidt RF (1962) Central pathways responsible for depolarization of primary afferent fibres. J Physiol (Lond) 160:237–257
Eccles JC, Schmidt RF, Willis WD (1963a) Pharmacological studies on presynaptic inhibition. J Physiol (Lond) 168:500–530
Eccles JC, Schmidt RF, Willis WD (1963b) Depolarization of central terminals of group Ib afferent fibres from muscle. J Neurophysiol 26:1–27
Eccles JC, Schmidt RF, Willis WD (1963c) The location and mode of action of the presynaptic inhibitory pathways on to group I afferent fibres from muscle. J Neurophysiol 26:506–522
Eccles JC, Llinás R, Sasaki K (1966) Parallel fibre stimulation and responses induced thereby in the Purkinje cells of the cerebellum. Exp Brain Res 1:17–39
Eckert R, Tillotson D (1978) Potassium activation associated with intraneuronal free calcium. Science 200:437–439
Eklof B, Siesjö BK (1972) The effect of bilateral carotid artery ligation upon acid-base parameters and substrate levels in the rat brain. Acta Physiol Scand 86:528–538
Endres W, Franz C, Nicholson C, ten Bruggencate G (1985) Stimulus-induced changes of extracellular pH in brain slice and peripheral nervous structures *in vitro*. Pflügers Arch [Suppl] 408:R44
Endres WK, Grafe P, Bostock H, ten Bruggencate G (1986a) Changes in extracellular pH during electrical stimulation of isolated rat vagus nerve. Neurosci Lett 64:201–205
Endres WK, Ballanyi K, Serve B, Grafe P (1986b) Excitatory amino acids and intracellular pH in motoneurons of the isolated spinal cord. Neurosci Lett 72:54–58

Ernst M, Lee MHM, Dworkin B, Zaretsky HH (1986) Pain perception decrement produced through repeated stimulation. Pain 26:221–231

Erulkar SD, Weight FF (1977) Extracellular potassium and transmitter release at the giant synapse of squid. J Physiol (Lond) 266:209–218

Evans MG, Thomas RC (1984) Acid influx into snail neurones caused by reversal of the normal pH_i – regulating system. J Physiol (Lond) 346:143–154

Fencl V (1986) Acid-base balance in cerebral fluids. In: The respiratory system. Control of breathing. American Physiological Society, Bethesda, pp 115–140 (Handbook of physiology, sect 3, vol II)

Fencl V, Miller TB, Pappenheimer JR (1966) Studies on the respiratory response to disturbances of acid-base balance, with deductions concerning the ionic composition of cerebral interstitial fluid. Am J Physiol 210:459–472

Fertzier AP, Ranck JB (1970) Potassium accumulation in interstitial space during epileptiform seizures. Exp Neurol 26:571–585

Fifková E, Bureš J (1964) Spreading depression in the mammalian striatum. Arch Int Physiol Biochim 72:171–179

Fisher RS, Pedley TA, Moody WJ Jr, Prince DA (1976a) The role of extracellular potassium in hippocampal epilepsy. Arch Neurol 33:76–83

Fisher RS, Pedley TA, Prince DA (1976b) Kinetics of potassium movement in normal cortex. Brain Res 101:223–237

Förstl J, Galvan M, Bruggencate ten G (1982) Extracellular K^+ concentration during electrical stimulation of rat isolated sympathetic ganglia, vagus and optic nerves. Neuroscience 7:3221–3229

Foster RE, Connors BM, Waxman SG (1982) Rat optic nerve: electrophysiological, pharmacological, and anatomical studies during development. Dev Brain Res 3:371–386

Frank K, Fuortes MGF (1957) Presynaptic and postsynaptic inhibition of monosynaptic reflexes. Fed Proc 16:39–40

Frankenhauser B, Hodgkin AL (1956) The after-effects of impulses in the giant nerve fibers of Loligo. J Physiol (Lond) 131:341–376

Frankenhauser B, Hodgkin AL (1957) The action of calcium on the electrical properties of squid axons. J Physiol (Lond) 137:218–244

Fritz LC, Gardner-Medwin AR (1976) The effect of synaptic activation on the extracellular potassium concentration in the hippocampal dentate areas in vitro. Brain Res 112:183–187

Frömter E, Burckhardt BC, Kondo Y (1988) Mechanisms of basolateral base transport in the renal proximal tubule. In: Thomas RC (ed) Proton passage across cell membranes. Wiley, Chichester, pp 106–121

Futamachi KJ, Mutani R, Prince DA (1974) Potassium activity in rabbit cortex. Brain Res 75:5–25

Gage PW, Quastel DMJ (1965) Dual effect of potassium on transmitter release. Nature 206:625–626

Gallagher JP, Higashi H, Nishi S (1978) Characterization and ionic basis of GABA-induced depolarization recorded in vitro from rat primary afferent neurones. J Physiol (Lond) 275:263–282

Gardner-Medwin AR (1983a) A study of the mechanisms by which potassium moves through brain tissue in the rat. J Physiol (Lond) 335:353–374

Gardner-Medwin AR (1983b) Analysis of potassium dynamics in mammalian brain tissue. J Physiol (Lond) 335:393–426

Gardner-Medwin AR, Gibson J, Willshaw DJ (1979) The mechanism of potassium dispersal in brain tissue. J Physiol (Lond) 293:37P–38P

Gardner-Medwin AR, Coles JA, Tsacopoulos M (1981) Clearance of extracellular potassium: evidence for spatial buffering by glial cells in the retina of the drone. Brain Res 209:452–457

Gasser HS, Graham HT (1933) Potentials produced in the spinal cord by stimulation of the dorsal roots. Am J Physiol 103:303–320
Giacobini E (1962) A cytochemical study of the localization of carbonic anhydrase in the nervous system. J Neurochem 9:169–177
Gilmore SA (1971) Neuroglial population in the spinal white matter of neonatal and early postnatal rats: an autoradiographic study of numbers of neuroglia and changes in their proliferative activity. Anat Rec 171:283–292
Gilmore SA, Sims TJ, Heard JK (1982) Autoradiographic and ultrastructural studies of areas of spinal cord occupied by Schwann cell myelin. Brain Res 239:365–375
Glötzner FL (1973) Membrane properties of neuroglia in epileptogenic gliosis. Brain Res 55:159–171
Gluck S Caldwell J (1988) Proton-translocating ATPase from bovine kidney medulla: partial purification and reconstitution. Am J Physiol 254 (Renal Fluid Electrolyte Physiol 23):F71–F79
Glynn IM Karlish SJD (1975) The sodium pump. Annu Rev Physiol 37:13–55
Grafe P, Reddy MM, Emmert H, Bruggencate ten G (1983) Effects of lithium on electrical activity and potassium ion distribution in the vertebrate central nervous system. Brain Res 279:65–76
Grafstein B (1956) Mechanism of spreading cortical depression. J Neurophysiol 19:154–171
Gray EG (1962) A morphological basis for presynaptic inhibition? Nature 193:82–83
Gray EG (1963) Electron microscopy of presynaptic organelles of the spinal cord. J Anat 97:101–106
Gray PTA, Ritchie JM (1986) A voltage-gated chloride conductance in rat cultured astrocytes. Proc R Soc Lond B228:267–288
Green JD (1964) The hippocampus. Physiol Rev 44:561–608
Grinstein S, Cohen S, Rothstein A (1984) Cytoplasmic pH regulation in thymic lymphocytes by an amiloride-sensitive Na^+/H^+ antiport. J. Gen Physiol 83:341–369
Grossman Y, Spira ME, Parnas I (1973) Differential flow of information into branches of a single axon. Brain Res 64:379–386
Gruol DL, Barker JL, Huang LM, McDonald JF, Smith TG Jr (1980) Hydrogen ions have multiple effects on the excitability of cultured mammalian neurons. Brain Res 183:247–252
Gutnick MJ, Connors BW, Ransom B (1981) Dye-coupling between glial cells in the guinea pig neocortical slice. Brain Res 213:486–492
Hablitz JJ, Heinemann U (1987) Extracellular K^+ and Ca^{2+} changes during epileptiform discharges in the immature rat neocortex. Dev Brain Res 36:299–303
Hablitz JJ, Heinemann U (1989) Alterations in the microenvironment during spreading depression associated with epileptiform activity in the immature neocortex. Dev Brain Res 76:243–252
Hablitz JJ, Lundervold A (1981) Hippocampal excitability and changes in extracellular potassium. Exp Neurol 71:410–420
Hackman JC, Ryan GP, Davidoff RA (1982) Excitatory amino acids and potassium release in the frog spinal cord. Neurosci Lett 33:289–293
Hagins WA, Penn RD, Yoshikami S (1970) Dark current and photocurrent in retinal rods. Biophys J 143:3802–412
Hagiwara S, Tasaki I (1958) A study on the mechanism of impulse transmission across the giant synapse of the squid. J Physiol (Lond) 143:114–137
Hájek I, Syková E (1981) Potassium-induced decrease of enkephalin binding in the frog spina cord. Physiol Bohemoslov 30:428–429
Hájek I, Teisinger J, Syková E (1985) The effect of opioid and of naloxone on Na^+, K^+-adenosine triphosphatase activity in frog spinal cord membrane fractions. Neurosci Lett 59:291–295

Hájek I, Svoboda J, Motin V, Syková E (1988) Extracellular potassium and pH changes in rat spinal cord evoked by nociceptive stimulation and peripheral injury. In: Abstracts of the symposium on integrative action of neuron: molecular basis, Moskva, p 37

Handwerker HO (1980) Pain producing substances. In: Kosterlitz HW, Terenius LY (eds) Pain and society, Chemie, Weinheim. pp 325–337

Hanitzsch R (1973) Intraretinal isolation of P III subcomponents in the isolated rabbit retina after treatment with sodium aspartate. Vision Res 13:2093–2102

Hansen AJ (1977) Extracellular potassium concentration in juvenile and adult rat brain cortex during anoxia. Acta Physiol Scand 99:412–420

Hansen AJ, Zeuthen T (1980) Brain extracellular space during spreading depression and ischemia. Acta Physiol Scand 108:355–365

Hansen AJ, Zeuthen T (1981) Extracellular ion concentrations during spreading depression and ischemia in rat brain cortex. Acta Physiol Scand 113:437–445

Hansen AJ, Lund-Andersen H, Crone C (1977) K^+-permeability of the blood-brain barrier, investigated by aid of a K^+-sensitive microelectrode. Acta Physiol Scand 101:438–445

Hansen AJ, Gjedde A, Siemkowicz E (1980) Extracellular potassium and blood flow in the post-ischemic rat brain. Pflugers Arch 389:1–7

Harris RJ, Symon L (1984) Extracellular pH, potassium, and calcium activities in progressive ischemia of rat cortex. J Cereb Blood Flow Metab 4:178–186

Harris RJ, Wieloch T, Symon L, Siesjö BK (1984) Cerebral extracellular calcium activity in severe hypoglycemia: relation to extracellular potassium and energy state. J Cereb Blood Flow Metab 4:187–193

Harris RJ, Richards PG, Symon L, Habib A-HA, Rosenstein J (1987) pH, K^+, and pO_2 of the extracellular space during ischaemia of primate cerebral cortex. J Cereb Blood Flow Metab 7:599–604

He L (1987) Involvement of endogenous opioid peptides in acupuncture analgesia. Pain 31:99–121

Heinemann U, Dietzel I (1984) Extracellular potassium concentration in chronic alumina cream foci of cats. J Neurophysiol 3:421–434

Heinemann U, Louvel J (1983) Changes in $[Ca^{2+}]_o$ and $[K^+]_o$ during repetitive electrical stimulation and during pentetrazole induced seizure activity in the sensorimotor cortex of rats. Pflügers Arch 398:310–317

Heinemann U, Lux HD (1975) Undershoots following stimulus-induced rises of extracellular potassium concentration in the cerebral cortex of cat. Brain Res 93:63–76

Heinemann U, Lux HD (1977) Ceiling of stimulus-induced rises in extracellular potassium concentration in the cerebral cortex of cat. Brain Res 120:231–249

Heinemann U, Lux HD, Gutnick MJ (1977) Extracellular free calcium and potassium during paroxysmal activity in the cerebral cortex of the cat. Exp Brain Res 27:237–243

Heinemann U, Konnerth A, Lux HD (1978) Changes in extracellular free Ca^{2+} and K^+ activity in epileptogenic alumina cream focus in the cerebral cortex of cats. Neurosci Lett [Suppl] 1:63

Heinemann U, Konnerth A, Lux HD (1981) Stimulation induced changes in extracellular free calcium in normal and chronic alumina cream foci of cats. Brain Res 213:246–250

Heinemann U, Neuhaus S, Dietzel I (1983) Aspects of potassium regulation in normal and gliotic brain tissue. In: Moulinier M, Meldrum BS (eds) Cerebral blood flow, metabolism and epilepsy. Libbey, London, pp 271–278

Heinemann U, Schaible H-G, Schmidt RF (1990) Changes in extracellular potassium concentration in cat spinal cord in response to innoxious and noxious stimulation of legs with healthy and inflamed knee joints. Exp Brain Res 79:283–292

Henderson LJ (1908) The theory of neutrality regulation in the animal organism. Am J Physiol 21:427–448

Henn FA, Haljamäe H, Hamberger A (1972) Glial cell function: Active control of extracellular K^+ concentration. Brain Res 43:437–443
Hertz L (1965) Possible role of neuroglia: a potassium-mediated neuronal-neuroglial-neuronal impulse transmission system. Nature 206:1091–1094
Hertz L (1977) Drug-induced alterations of ion distribution at the cellular level of the central nervous system. Pharmacol Rev 29:35–65
Hertz L (1978) An intense potassium uptake into astrocytes, its further enhancement by high potassium, and its possible involvement in potassium homeostasis at the cellular level. Brain Res 145:202–208
Hertz, L. (1979) Functional interactions between neurons and astrocytes. I. Turnover and metabolism of putative amino-acid transmitters. Prog Neurobiol 13:277–323
Hertz L, Schousboe A (1975) Ion and energy metabolism of the brain at the cellular level. Int Rev Neurobiol 18:141–211
Heuser D, Hossmann KA, Schindler U, Betz E (1975) Changes in extracellular ion activities and brain volume during prolonged cerebral ischaemia and recovery. Pflugers Arch 355 [Suppl 198]:R99
Hiedl P, Struppler A, Gessler M (1979) Local analgesia by percutaneous electrical stimulation of sensory nerves. Pain 7:129–134
Hille B (1968) Charges and potentials at the nerve surface: divalent ions and pH. J Gen Physiol 51:221–236
Hinke JAM (1959) Glass microelectrodes for measuring intracellular activities of sodium and potassium. Nature 184:1257–1258
Hösli E, Hösli L (1976) Uptake of L-glutamate and L-aspartate in neurones and glial cells of cultured human and rat spinal cord. Experientia 32:219–222
Holobut W, Niechaj A (1973) The dorsal root potentials produced on both sides of the spinal cord by longlasting stimulation of the cutaneous afferents. J Physiol (Lond) 230:15–27
Hotson JR, Sypert GW, Ward AA Jr (1973) Extracellular potassium concentration changes during propagated seizures in neocortex. Exp Neurol 38:20–26
Howland B, Lettwin JY, McCulloch WS, Pitts W, Wall PD (1955) Reflex inhibition by dorsal root interaction. J Neurophysiol 18:1–17
Hubbard JI, Willis WD (1962) Hyperpolarization of mammalian motor nerve terminals. J Physiol (Lond) 163:115–137
Hughes JR (1958) Post-tetanic potentiation. Physiol Rev 38:91–113
Hunskaar S, Fasmer OB, Hole K (1985) Formalin test in mice, a useful technique for evaluation mild analgesics. J Neurosci Methods 14:69–76
Husted RF, Reed DJ (1976) Regulation of cerebrospinal fluid potassium by cat choroid plexus. J Physiol (Lond) 259:213–221
Huxley AF, Stämpfli R (1951) Effect of potassium and sodium on resting and action potentials of single myelinated nerve fibers. J Physiol (Lond) 112:496–508
Iljima T, Ciani S, Hagiwara S (1986) Effects of the external pH on Ca channels: Experimental studies and theoretical considerations using a two-site, two-ion model. Proc Natl Acad Sci USA 83:654–658
Ito U, Go KG, Walker JT Jr, Spatz M, Klatzo I (1976) Experimental cerebral ischemia in mongolian gerbils. III. Behaviour of the blood-brain barrier. Acta Neuropathol (Berl) 34:1–6
Iwasa K, Tasaki I, Gibbons RC (1980) Swelling of nerve fibers associated with action potentials. Science 210:338–339
Izquierdo I (1972) Hippocampal physiology: experiments on regulation of its electrical activity, on the mechanism of seizures, and on a hypothesis of learning. Behav Biol 7:669–698
Izquierdo JA, Izquierdo I (1967) Electrolytes and excitable tissues. Annu Rev Pharmacol 7:125–144
Izquierdo I, Nasello AG, Marichich ES (1971) The dependence of hippocampal function on extracellular potassium levels. Curr Mod Biol 4:35–46

Javaheri S, DeHemptinne A, Vanheel B, Leusen I (1983) Changes in brain ECF pH during metabolic acidosis and alkalosis: a microelectrode study. J Appl Physiol 55(6):1849–1853

Jendelová P, Chvátal A, Syková E (1990) Alkaline-acid changes in the extracellular pH in the frog spinal cord caused by H^+ fluxes through proton channels. Physiol Bohemoslov 39:270

Jendelová P, Syková E (1991) Role of glia in K^+ and pH homeostasis in the neonatal rat spinal cord. Glia 4:56–63

Jendelová P, Ng KT, Chvátal A, Syková E (1988) Extracellular K^+ concentration in the chick brain and cerebrospinal fluid. Physiol Bohemoslov 37:516

Johanson CE, Reed DJ, Woodbury DM (1974) Active transport of sodium and potassium by the choroid plexus of the rat. J Physiol (Lond) 241:359–372.

Johnson JH, Dunn DP, Rosenberg RN (1982) Furosemide-sensitive K^+ channel in glioma cells but not neuroblastoma cells in culture. Biochem Biophys Res Commun 109:100–115

Johnstone BM, Patuzzi R, Syka J, Syková E (1989) Stimulus-releated potassium changes in the organ of Corti of guinea-pig. J Physiol (Lond) 408:77–92

Johnstone C, Schmidt R, Johnstone B (1963) Sodium and potassium in vertebrate cochlear endolymph as determined by flame microphotometry. Comp Biochem Physiol 9:33–341

Kaila K, Vaughan-Jones RD (1987) Influence of sodium-hydrogen exchange on intracellular pH, sodium and tension in sheep cardiac Purkinje fibres. J Physiol (Lond) 390:93–118

Kaila K, Voipio J (1987) Postsynaptic fall in intracellular pH induced by GABA-activated bicarbonate conductance. Nature 330:163–165

Karwoski CJ, Proenza LM (1987) Sources and sinks of light-evoked $\Delta[K^+]_o$ in the vertebrate retina. Can J Physiol Pharmacol 65:1009–1017

Karwoski CJ, Newman EA, Shimazaki H, Proenza LM (1985) Light-evoked increases in extracellular K^+ in the plexiform layers of amphibian retinas. J Gen Physiol 86:189–213

Karwoski CJ, Lu H-K, Newman EA (1989) Spatial buffering of light-evoked potassium increases by retinal Müller (glial) cells. Science 244:578-580

Kasamatsu T (1970) Maintained and evoked unit activity in the mesencephalic reticular formation of the freely behaving cat. Exp Neurol 28:450–470

Katchalsky AK, Rowland V, Blumental R (1974) Dynamic patterns of brain cell assemblies. Neurosci Res Prog Bull 12:3–187

Katz B (1966) Nerve, muscle and synapse. McGraw-Hill, New York, pp 1–193

Katzman R, Pappius HM (1973) Brain electrolytes and Fluid metabolism. William and Wilkins, Baltimore

Kemény A, Boldizsár H, Pethes G (1961) The distribution of cations in plasma and cerebrospinal fluid following infusion of solutions of salts of sodium, potassium, magnesium, and calcium. J Neurochem 7:218–227

Kempski O (1986) Cell swelling mechanisms in brain. In: Baethmann A, Go KG, Unterberg A (eds) Mechanisms of secondary brain damage. Plenum, New York, pp 203–220

Kempski O, Gross U, Baethman A (1982) An in vitro model of cytotoxic brain edema: cell volume and metabolism of cultivated glial and nerve cells. In: Driesen W, Brock M, Klinger M (eds) Computerized tomography. Brain metabolism. Spinal injuries. Springer, Berlin Heidelberg New York (Advances in Neurosurgery, vol 10)

Kempski O, Chaussy L, Gross U, Zimmer M, Baethmann A (1983) Volume regulation and metabolism of suspended C6 glioma cells: an in vitro model to study cytotoxic brain edema. Brain Res 279:217–228

Kettenmann H, Schlue WR (1988) Intracellular pH regulation in cultured mouse oligodendrocytes. J Physiol (Lond) 406:147–162

Kettenmann H, Orkand RK, Lux HD, Schachner M (1982) Single potassium channel currents in cultured mouse oligodendrocytes. Neurosci Lett 32:41–46

Kettenmann H, Orkand RK, Schachner M (1983a) Coupling among identified cells in nervous system cultures. J Neurosci 3:506–516

Kettenmann H, Sonnhof U, Schachner M (1983b) Exclusive potassium dependence of the membrane potential in cultured mouse oligodendrocytes. J Neurosci 3:500–505

Kettenmann H, Syková E, Orkand RK, Schachner M (1987) Glial potassium uptake following depletion by intracellular ionophoresis. Pflügers Arch 410:1–6

Keynes RD (1951) The leakage of radioactive potassium from stimulated nerve. J Physiol (Lond) 113:99–114

Keynes RD, Lewis PR (1951) The sodium and potassium content of cephalopod nerve fibres. J Physiol (Lond) 114:151–182

Keynes RD, Ritchie JM (1965) The movements of labelled ions in mammalian non-myelinated nerve fibres. J Physiol (Lond) 179:333–367

Kimelberg HK (1979) Glial enzymes and ion transport in brain swelling. In: Popp AJ, Bourke RS, Nelson LR, Kimelberg HK (eds) Neural trauma Raven, New York, pp 137–153

Kimelberg HK (1981) Active accumulation and exchange transport of chloride in astroglial cells in culture. Biochim Biophys Acta 646:179–184

Kimelberg HK, Frangakis MV (1985) Furosemide and bumetanide sensitive ion transport and volume control in primary astrocyte cultures from rat brain. Brain Res 361:125–134

Kimelberg HK, Bidlecome S, Narumi S, Bourke RS (1978) ATPase and carbonic anhydrase activities of bulk-isolated neuron, glia and synaptosome fractions from rat brain. Brain Res 141:305–323

Kimelberg HK, Biddlecome S, Bourke RS (1979a) SITS-inhibitable Cl^- transport and Na^+-dependent H^+ production in primary astroglial cultures. Brain Res 173:111–124

Kimelberg HK, Bowman CL, Biddlecome S, Bourke RS (1979b) Cation transport and membrane potential properties of primary astroglial cultures from neonatal rat brains. Brain Res 177:533–550

Kirshner HS, Blank WF, Myers RE (1976) Changes in cortical subarachnoid fluid potassium concentrations during hypoxia. Arch Neurol 33:84–90

Klatzo I (1967) Neuropathological aspects of brain edema. J Neuropathol Exp Neurol 26:1–14

Kline RP, Ripps H, Dowling JE (1978) Generation of b-wave currents in the skate retina. Proc Natl Acad Sci USA 75:5727–5731

Kocsis JD, Malenka RC, Waxman SG (1981) Enhanced parallel fiber frequency-following after reduction of postsynaptic activity. Brain Res 207:321–331

Kocsis JD, Malenka RC, Waxman SG (1983) Effects of extracellular potassium concentration on the excitability of the parallel fibres of the rat cerebellum. J Physiol (Lond) 334:225–244

Koketsu K (1956a) Intracellular potential changes of primary afferent nerve fibers in spinal cords of cats. J Neurophysiol 19:375–392

Koketsu K (1956b) Intracellular slow potential of dorsal root fibers. Am J Physiol 184:338–344

Konnerth A, Heinemann U, Yaari Y (1986) Nonsynaptic epileptogenesis in the mammalian hippocampus in vitro I. Development of seizure like activity in low extracellular calcium, J. Neurophysiol 56:409–423

Konnerth A, Lux HD, Morad M (1987) Proton induced transformation of calcium channel in chicks dorsal root ganglion cells. J Physiol (Lond) 386:603–633

Koroleva VI, Bureš J (1979) Circulation of cortical spreading depression around electrically stimulated areas and epileptic foci in the neocortex of rats. Brain Res 173:209–215

Koroleva VI, Bureš J (1980) Blockade of cortical spreading depression in electrically and/or chemically stimulated areas of cerebral cortex in rats. EEG Clin Neurophysiol 48:1–15

Koroleva VI, Bureš J (1982) Stimulation induced recurrent epileptiform discharges block cortical and subcortical spreading depression in rats. Physiol Bohemoslov 31:385–400

Koryta J (1980) Medical and Biological Applications of Electrochemical Devices. Wiley, Chichester

Kosterlitz HW, McKnight AT (1981) Opioid peptides and sensory function. In: Ottoson D (ed) Progress in sensory physiology, vol I. Springer, Berlin Heidelberg New York, pp 31–95

Kraig RP, Nicholson C (1976) Sodium liquid ion exchanger microelectrode used to measure large extracellular sodium transients. Science 194:725–726

Kraig RP, Nicholson C (1978) Extracellular ionic variations during spreading depression. Neuroscience 3:1045–1059

Kraig RP, Petito CK (1989) Interrelation of proton and volume regulation in astrocytes. In Ginsberg MD, Dietrich WD (eds) Cerebrovascular diseases. Raven, New York, pp 239–246

Kraig RP, Ferreira-Filho CR, Nicholson C (1983) Alkaline and acid transients in cerebellar microenvironment. J Neurophysiol 49:831–850

Kraig RP, Pulsinelli WA, Plum F (1985a) Heterogenous distribution of hydrogen and bicarbonate ions during complete brain ischemia. Prog Brain Res 63:155–166

Kraig RP, Pulsinelli WA, Plum F (1985b) Hydrogen ion buffering during complete brain ischemia. Brain Res 342:281–290

Kraig RP, Pulsinelli WA, Plum F (1986) Carbonic acid buffer changes during complete brain ischemia. Am J Physiol 250:R348–R357

Kraig RP, Petito CK, Plum F, Pulsinelli WA (1987) Hydrogen ions kill brain at concentrations reached in ischemia. J Cereb Blood Flow Metab 7:379–386

Krishtal OA, Pidoplichko VI (1980) A receptor for protons in the nerve cell membrane. Neuroscience 5:2325–2327

Krishtal OA, Pidoplichko VI (1981a) A receptor for protons in the membrane of sensory neurons may participate in nociception. Neuroscience 6:2599–2601

Krishtal OA, Pidoplichko VI (1981b) A "receptor" for protons in small neurons of trigeminal ganglia: possible role in nociception. Neurosci Lett 24:243–246

Krishtal OA, Osipchuk YV, Shelest TN, Smirnov SV (1987) Rapid extracellular pH transients related to synaptic transmission in rat hippocampal slices. Brain Res 436:352–356

Křivánek J, Bureš J (1960) Ion shifts during Leão's spreading cortical depression. Physiol Bohemoslov 9:494–503

Křivánek J, Reddy MM (1980) Does vanadate participate in the response of cerebral $[Na^{+}$-$K^{+}]$-ATPase to electrical stimulation? Brain Res 195:506–510

Kříž N, Syková E (1981) Sensitivity of K^{+}-selective microelectrodes to pH and some biologically active substances. In: Syková E, Hník P, Vyklický (eds) Ion-selective microelectrodes and their use in excitable tissues. Plenum, New York, pp 25–39

Kříž N, Syková E, Ujec E, Vyklický L (1974) Changes of extracellular potassium concentration induced by neuronal activity in the spinal cord of the cat. J Physiol (Lond) 238:1–15

Kříž N, Syková E, Vyklický L (1975) Extracellular potassium changes in the spinal cord of the cat and their relation to slow potentials, active transport and impulse transmission. J Physiol (Lond) 249:167–182

Krnjevič K (1974) Chemical nature of synaptic transmission in vertebrates. Physiol Rev 54:418–540

Krnjevič K, Morris ME (1972) Extracellular K^{+} activity and slow potential changes in spinal cord and medulla. Can J Physiol Pharmacol 50:1214–1217

Krnjevič K, Morris ME (1974) Extracellular accumulation of K^{+} evoked by activity of

primary afferent fibers in the cuneate nucleus and dorsal horn of cats. Can J Physiol Pharmacol 52:352–871
Krnjevič K, Morris ME (1975) Factors determining the decay of K^+ potentials and focal potentials in the central nervous system. Can J Physiol Pharmacol 53:923–934
Krnjevič K, Morris ME (1976) Input-output relation of transmission through cuneate nucleus. J Physiol (Lond) 257:791–815
Krnjevič K, Morris ME (1981) Electrical and functional correlates of changes in transmembrane ionic gradients produced by neural activity in the central nervous system. In: Zeuthen T (ed) The application of ion-selective microelectrodes. Elsevier, Amsterdam, pp 165–216
Krnjevič K, Morris ME, Reiffenstein RJ (1980) Changes in extracellular Ca^{2+} and K^+ activity accompanying hippocampal discharges. Can J Physiol Pharmacol 58:579–583
Krnjevič K, Morris ME, Reiffenstein RJ (1982a) Stimulation-evoked changes in extracellular K^+ and Ca^{2+} in pyramidal layers of the rat's hippocampus. Can J Physiol Pharmacol 60:1643–1657
Krnjevič K, Morris ME, Reiffenstein RJ, Ropert N (1982b) Depth distribution and mechanism of changes in extracellular K^+ and Ca^{2+} concentrations in the hippocampus. Can J Physiol Pharmacol 60:1658–1671
Kudo Y, Fukuda H (1976) Alteration of extracellular K^+ activity induced by amino acids in frog spinal cord. Jpn J Pharmacol 26:385–387
Kuffler SW (1967) Neuroglial cells: physiological properties and a potassium mediated effect of neuronal activity on the glial membrane potential. Proc R Soc Lond [B] 168:1–21
Kuffler SW, Nicholls JG (1966) The physiology of neuroglial cells. Ergeb Physiol Biol Chem Exp Pharmacol 57:1–90
Kuffler SW, Nicholls JG, Orkand RK (1966) Physiological properties of glial cells in the central nervous system of amphibia. J Neurophysiol 29:768–787
Kusano K, Livengood DR, Werman R (1967) Correlation of transmitter release with membrane properties of the presynaptic fiber of the squid giant synapse. J Gen Physiol 50:2579–2601
Kuschinsky W, Wahl M (1978) Local chemical and neurogenic regulation of central vascular resistance. Physiol Rev 58:656–689
Laming PR (1989) Do glia contribute to behaviour? A neuromodulatory review. Comp Biochem Physiol 94A:555–568
Leão AAP (1944) Spreading depression of activity in the cerebral cortex. J Neurophysiol 7:359–390
Lehmenkühler A, Zidek W, Staschen M, Caspers H (1981) Cortical pH and pCa in relation to DC potential shifts during spreading depression and asphyxiation. In: Syková E, Hník P, Vyklický L (eds) Ion-selective microelectrodes and their use in excitable tissues. Plenum, New York, pp 225–229
Levin VA, Fenstermacher JD, Patlak CS (1970) Sucrose and inulin space measurements of cerebral corkex in four mammalian species. Am J Physiol 219:1528–1533
Levy RA (1977) The role of GABA in primary afferent depolarization. Prog Neurobiol 9:211–267
Levy RA (1980) Presynaptic control of input to the central nervous system. Can J Physiol Pharmacol 58:751–766
Lewis DV, Schuette WK (1975a) Temperature dependence of potassium ion clearance in the nervous system. Brain Res 99:175–178
Lewis DV, Schuette WH (1976) NADH fluorescence, $[K^+]_o$ and oxygen consumption in cat cerebral cortex during direct cortical stimulation. Brain Res 110:523–535
Lewis DV, Matsuga N, Schuette WH, Van Buren J (1977) Potassium clearance and reactive gliosis in the alumina gel lesion. Epilepsia 18:499–506
Liley AW (1956) The effect of presynaptic polarization on the spontaneous activity at the mammalian neuromuscular junction. J Physiol (Lond) 134:427–443

Linser P, Moscona AA (1984) Variable CA II compartmentalization in vertebrate retina. Ann NY Acad Sci 429:430–446
Lipton P (1973) Effect of membrane depolarization on light scattering by cerebral cortical slices. J Physiol (Lond) 231:365–382
Lipton P, Heimbach CJ (1977) The effect of extracellular potassium concentration on protein synthesis in guinea-pig hippocampal slices. J Neurochem 28:1347–1354
Lloyd DPC, McIntyre AK (1949) On the origin of dorsal root potentials. J Gen Physiol 32:409–443
Long DM, Hagfors N (1975) Electrical stimulation in the nervous system: the current status of electrical stimulation in the nervous system for relief of pain. Pain 1:109–123
Lothman E, LaManna JC, Cordingley G, Rosenthal M, Somjen G (1975) Responses of electrical potential, potassium levels and oxidative metabolic activity of the cerebral neocortex of cats. Brain Res 88:15–36
Lothman EW, Somjen GG (1975) Extracellular potassium activity, intracellular and extracellular potential responses in the spinal cord. J Physiol (Lond) 252:115–136
Lothman EW, Somjen GG (1976) Functions of primary afferents and responses of extracellular K^+ during spinal epileptiform seizures. Electroencephalogr Clin Neurophysiol 41:253–267
Lux HD (1974) The kinetics of extracellular potassium: relation to epileptogenesis. Epilepsia 15:375–393
Lux HD, Neher E (1973) The equilibration time course of $[K^+]_o$ in cat cortex. Exp Brain Res 17:190–205
Lynch GS, Gribkoff VK, Deadwyler SA (1976) Long term potentiation is accompanied by a reduction in dendritic responsiveness to glutamic acid. Nature 263:151–153
Mac Gregor RJ, Prieto-Diaz R, Miller SW, Groves RM (1973) Statistical properties of neurons in the rat mesencephalic reticular formation. Brain Res 64:167–187
MacKnight ADC, Leaf A (1977) Regulation of cellular volume. Physiol Rev 37:510–573
MacVicar BA (1984) Voltage-dependent calcium channels in glial cells. Science 226:1345–1347
Malenka RC, Kocsis JD (1982) Effects of GABA on stimulus-evoked changes in $[K^+]_o$ and parallel fiber excitability. J Neurophysiol 48:608–621
Malenka RC, Kocsis JD, Ransom BR, Waxman SG (1981) Modulation of parallel fiber excitability by post-synaptically mediated changes in extracellular potassium. Science 214:339–341
Malenka RC, Kauer JA, Perkel DJ, Nicoll RA (1989) The impact of postsynaptic calcium on synaptic transmission – its role in long-term potentiation. Trends Neurosci 12:444–450
Manohar S, Noda H, Adey RW (1972) Behavior of mesencephalic reticular neurons in sleep and wakefulness. Exp Neurol 34:140–157
Maren TH (1967) Carbonic anhydrase: chemistry, physiology and inhibition. Physiol Rev 47:595–781
Margolis RU, Margolis RK (1977) Metabolism and function of glycoproteins and glycosaminoglycans in nervous tissue. Int J Biochem 8:85–91
Matsuura S (1969) Effects of changes in extracellular K^+ concentration on resting and action potentials of the toad spinal motoneuron. Osaka City Med J 15:29–45
Mayevsky A, Zeuthen T, Chance B (1974) Measurements of extracellular potassium, ECoG and pyridine nucleotide levels during cortical spreading depression in rats. Brain Res 76:347–349
Meech RW, Thomas RC (1987) Voltage-dependent intracellular pH in Helix aspersa neurones. J Physiol (Lond) 390:433–452
Melichar I, Syka J (1977) Time course of anoxia-induced K^+ concentration changes in the cochlea measured with K^+ specific microelectrodes. Pflügers Arch 372:207–213

Melichar I, Syka J (1987) Electrophysiological measurements of the stria vascularis potentials in vivo. Hear Res 25:35–43
Mellerová B, Syková E (1978) Sensitivity of dorsal root potentials to changes of extracellular potassium. Physiol Bohemoslov 27:261–262
Melzack R (1975) Prolonged relief of pain by brief, intense transcutaneous somatic stimulation. Pain 1:357–373
Melzack R, Wall PD (1965) Pain mechanisms: a new theory. Science 150:971–979
Mendell LM (1966) Physiological properties of the unmyelinated fibre projection to the spinal cord. Exp Neurol 16:316–332
Meyer JS, Gotoh F, Tazaki Y, Hamaguchi K, Ishikawa S, Nouailhat F, Symon L (1962) Regional cerebral blood flow and metabolism in vivo. Arch Neurol 7:560–581
Minchin MCW, Iversen LL (1974) Release of $^{3}[H^{+}]$ gamma-aminobutyric acid from glial cells in rat dorsal root ganglia. J Neurochem 23:533–540
Moody WJ (1981) The ionic mechanism of intracellular pH regulation in crayfish neurones. J Physiol (Lond) 316:293–308
Moody W Jr (1984) Effects of intracellular H^+ on the electrical properties of excitable cells. Annu Rev Neurosci 7:257–278
Moody W, Futamachi KJ, Prince DA (1974) Extracellular potassium activity during epileptogenesis. Exp Neurol 42:248–263
Moolenaar WH, Tsein RY, Van der Saag PT, de Laat SW (1983) Na^+/H^+ exchange and cytoplasmic pH in the action of growth factors in human fibroblasts. Nature 304:645–648
Mori S, Miller WM, Tomita T (1976) Müller cell function during spreading depression in frog retina. Proc Natl Acad Sci USA 73:1351–1354
Morris ME (1974) Hypoxia and extracellular potassium activity in the guinea pig cortex. Can J Physiol Pharmacol 52:872–882
Morris ME (1978) Brain extracellular potassium and general anaesthetics. Can J Physiol Pharmacol 56:863–872
Morris ME (1981) Measurements of ion activity in the CNS: extracellular K^+ and Ca^{2+} in the hippocampus. In: Syková E, Hník P, Vyklický L (eds) Ion-Selective microelectrodes and their use in excitable tissues. Plenum, New York, pp 241–246
Musser GL, Rosen S (1973) Localization of carbonic anhydrase activity in the vertebrate retina. Exp Eye Res 15:105–119
Mutani R, Futamachi KJ, Prince DA (1974) Potassium activity in immature cortex. Brain Res 75:27–39
Mutch WAC, Hansen AJ (1984) Extracellular pH changes during spreading depression and cerebral ischemia: mechanisms of brain pH regulation. J Cereb Blood Flow Metab 4:17–27
Mutsuga N, Schuette WH, Lewis DV (1976) The contribution of local blood flow to the rapid clearance of potassium from the cortical extracellular space. Brain Res 116:431–436
Nattie EE (1977) Brain tissue potassium in normal and potassium depleted rats. Life Sci 21:1851–1856
Nemoto EM, Frinak S (1981) Brain tissue pH after global brain ischemia and barbiturate loading in rats. Stroke 12:77–82
Newman EA (1985a) Voltage-dependent calcium and potassium channels in retinal glial cells. Nature 317:809–811
Newman EA (1985b) Regulation of potassium levels by glial cells in the retina. Trends Neurosci 8:156–159
Newman EA (1985c) Membrane physiology of retinal glial (Müller) cells. J Neurosci 5:2225–2239
Newman EA (1986) High potassium conductance in astrocytes endfeet. Science 233:453–454
Newman EA (1987) Regulation of potassium levels by Müller cells in the vertebrate retina. Can J Physiol Pharmacol 65:1028–1034

Nicholson C (1979) Brain-cell microenvironment as a communication channel. In: Schmitt, FO, Worden FG (eds) The neuroscience: fourth study program. MIT Press, Cambridge, pp 457–476
Nicholson C (1980) Dynamics of the brain cell microenvironment. Neurosci Res Program Bull 18:177–322
Nicholson C, Phillips JM (1979) Diffusion of anions and cations in the extracellular microenvironment of the brain. J Physiol (Lond) 296:66P
Nicholson C, Phillips JM (1981) Ion diffusion modified by tortusity and volume fraction in the extracellular micro-environment of the rat cerebellum. J Physiol (Lond) 321: 225–257
Nicholson C, Rice ME (1988) Use of ion-selective microelectrodes and voltammetric microsensors to study brain cell microenvironment. In: Boulton AA, Baker GB, Walz W (eds) Neuromethods: the neuronal microenvironment. Humana, New York, pp 247–361
Nicholson C, Rice ME (1991) Diffusion of ions and transmitters in the brain cell microenvironment. In: Fuxe K, Agnati LF (eds) Volume transmission in the brain. Raven, New York (in press)
Nicholson C, Brugencate ten G, Senekowitsch R (1976) Large potassium signals and slow potentials evoked during aminopyridine or barium superfusion in cat cerebellum. Brain Res 113:606–610
Nicholson C, Brugencate ten G, Steinberg R, Stöckle H (1977) Calcium modulation in brain extracellular microenvironment demonstrated by ion-selective micropipette. Proc Natl Acad Sci USA 74:1287–1290
Nicholson C, Brugencate ten G, Stöckle H, Steinberg R (1978) Calcium and potassium changes in extracellular microenvironment of cat cerebellar cortex. J Neurophysiol 41:1026–1039
Nicholson C, Phillips JM, Gardner-Medwin AR (1979) Diffusion from and iontophoretic point source in the brain: role of tortuosity and volume fraction. Brain Res 169:580–584
Nicholson C, Phillips, JM, Tobias C, Kraig RP (1981) Extracellular potassium, calcium and volume profiles during spreading depression. In: Syková E, Hník P, Vyklický L (eds) Ion-selective microelectrodes and their use in excitable tissues. Plenum, New York, pp 211–223
Nicholson C, Kraig RP, Ferreira-Filho CR, Thompson P (1985) Hydrogen ion variations and their interpretation in the microenvironment of the vertebrate brain. In Kessler M, Harrison DK, Höper J (eds) Ion measurments in physiology and medicine. Springer, Berlin Heidelberg New York, pp 229–235
Nicoll RA (1979) Dorsal root potentials and changes in extracellular potassium in the spinal cord of the frog. J Physiol (Lond) 290:113–127
Nistri A, Corradetti R (1978) A comparison of the effects of GABA, 3-aminopropanesulphonic acid and imidazoleacetic acid on the frog spinal cord. Neuropharmacology 17:13–19
Nistri S, Minolta S, Karczmar AG (1974) Primary afferent neurones: the ionic mechanism of GABA-mediated depolarization. Neuropharmacology 13:215–219
Noda H, Adey RW (1973) Neuronal activity in the association cortex of the cat during sleep, wakefulness and anaesthesia. Brain Res 54:243–259
Nowak L, Ascher P, Berwald-Netter Y (1987) Ionic channels in mouse astrocytes in culture. J Neurosci 7:101–109
Nuccitelli R, Deamer DW (eds) (1982) Intracellular pH: its measurement, regulation, and utilization in cellular functions. Liss, New York
Nyberg F, Yaksh TL, Terenius L (1983) Opioid activity released from cat spinal cord by sciatic nerve stimulation. Life Sci 33 [Suppl I]:17–20
Oakley B, Green DG (1976) Correlation of light-induced changes in retinal extracellular potassium concentration with c-wave of the electroretinogram. J Neurophysiol 39:1117–1133

Oakley B, Wen R (1989) Extracellular pH in the isolated retina of the toad in darkness and during illumination. J Physiol (Lond) 419:353–378
Oehme M, Simon W (1976) Microelectrode for K^+ based on a neutral carrier and comparison of its characteristics with a cation exchanger sensor. Anal Chim Acta 86:21–25
Oehme M, Kessler M, Simon W (1976) Neutral carrier Ca^{2+} microelectrode. Chimia 30:204–206
Ogata N (1975) Ionic mechanisms of the depolarization shift in thin hippocampal slices. Exp Neurol 46:147–155
Ogata N, Hori N, Katsuda N (1976) The correlation between extracellular potassium concentration and hippocampal epileptic activity in vitro. Brain Res 110:371–375
Ookawa T, Bureš J (1975) Extracellular potassium shifts accompanying epileptic discharge induced in chicken hyperstriatum by systemic injection of metrazol. Brain Res 97:171–176
Orkand RK, Nicholls JG, Kuffler SW (1966) Effects of nerve impulses on the membrane potential of glial cells in the central nervous system of amphibia. J Neurophysiol 29:788–806
Orkand PM, Bracho H, Orkand RK (1973) Glial metabolism: alteration by potassium levels comparable to those during neural activity. Brain Res 55:467–471
Orkand RK, Orkand PM, Tang C-M (1981) Membrane properties of neuroglia in the optic nerve of Necturus. J Exp Biol 95:49–59
Orkand RK, Dietzel I, Coles JA (1984) Light-induced changes in extracellular volume in the retina of the drone, Apis mellifera. Neurosci Lett 45:273–278
Pappius HM (1974) Fundamental aspects of brain edema. Vinkel PJ, Bruyn GW (eds) In: Handb für Klinische Neurology, vol 16, part I. Elsevier, Amsterdam
Patel SN, Stewart MG (1988) Changes in the number and structure of dendritic spines, 25 hours after passive avoidance training in the domestic chick, Gallus domesticus. Brain Res 449:331–335
Patlak CS, Fenstermacher JD (1975) Measurements of dog blood-brain transfer constants by ventriculocisternal perfusion. Am J Physiol 229:877–884
Pedley TA, Fischer RS, Futamachi KJ, Prince DA (1976a) Regulation of extracellular potassium concentration in epileptogenesis. Fed Proc 35:1254–1259
Pedley TA, Fischer RS, Prince DA (1976b) Focal gliosis and potassium movement in mammalian cortex. Exp Neurol 50:346–361
Pentreath VW (1982) Potassium signalling of metabolic interactions between neurons and glial cells. Trends Neurosci 10:339–345
Pentreath VW, Kai-Kai MA (1982) Significance of the potassium signal from neurones to glial cells. Nature 295:59–61
Phillips JM, Nicholson C (1978) Tetra-alkyl ammonium ions as probes of brain cell microenvironment. Soc Neurosci Abstr 4:236
Phillips JM, Nicholson C (1979) Anion permeability in spreading depression investigated with ion-sensitive microelectrodes. Brain Res 173:567–571
Pollen DC, Trachtenberg MC (1970) Neuroglia: gliosis and focal epilepsy. Science 167:1253–1258
Poolos NP, Mauck MD, Kocsis JD (1987) Activity-evoked increases in extracellular potassium modulate presynaptic excitability in the CA 1 region of the hippocampus. J Neurophysiol 2:404–416
Prince DA (1978) Neurophysiology of epilepsy. Annu Rev Neurosci 1:395–415
Prince DA, Schwartzkroin PA (1978) Non-synaptic mechanisms in epileptogenesis. In: Chalazonitis N, Boisson M (eds) Abnormal neuronal discharges. Raven, New York, pp 1–12
Prince DA, Lux HD, Neher E (1973) Measurement of extracellular potassium activity in cat cortex. Brain Res 50:489–495
Prince DA, Benninger C, Kadis J (1981) Evoked ionic alterations in brain slices. In: Syková E, Hník P, Vyklický L (eds) Ion-selective microelectrodes and their use in excitable tissues. Plenum, New York, pp 247–259

Pumain R, Kurcewicz I, Louvel J (1983) Fast extracellular calcium transients: involvement in epileptic processes. Science 222:177–179
Quandt FN, MacVicar BA (1986) Calcium activated potassium channels in cultured astrocytes. Neuroscience 19:29–41
Radil-Weiss T, Syková E (1977a) O vozmožnoj roli i vozniknovenii nevrozov vzaimootnošeni j vegetativnych processov i aktivnosti retikuljarnych nejronov (The possible role of the interrelationship between impulse activity of reticular neurons and vegetative processes in neurons) Zh Vyssh Nerv Deiat 22:322–324
Radil-Weiss T, Syková E (1977b) Vegetative Prozesse und die Aktivität retikulärer Neuronemögliche Beziehungen zur Entstehung von Neurosen. In: Baumann R, Hecht K (eds) Stress, Neurose und Herz-Kreislauf, Deutscher Verlag der Wissenschaften, Berlin, pp 109–113
Rall W (1974) Dendritic spines, synaptic potency and neuronal plasticity. In: Woody CD, Brown KA, Crow TJ Ir, Knispel JD (eds) Cellular mechanisms subserving changes in neuronal activity. Los Angeles, pp 13–22 (Brain information service-report, vol 3)
Ransom BR, Goldring S (1973) Ionic determinants of membrane potential of cells presumed to be glia in cerebral cortex of cat. J Neurophysiol 36:855–868
Ransom BR, Carlini WG, Connors B (1985a) Brain extracellular space: developmental studies in rat optic nerve. Ann NY Acad Sci 481:87–105
Ransom BR, Yamate CL, Connors BW (1985b) Activity dependent shrinkage of extracellular space in rat optic nerve: a developmental study. J Neurosci 5:532–535
Rapoport SI (1976) The blood-brain barrier in physiology and medicine. Raven, New York
Reddy MM, Bureš J (1980) Cortical $[K^+]_e$ and the stimulation induced blockade of spreading depression in the rat cerebral cortex. Neurosci Lett 17:243–247
Rehncrona S, Rosen I, Siesjö BK (1980) Excessive cellular acidosis: an important mechanism of neuronal damage in the brain. Acta Physiol Scand 110:45–437
Rexed B (1954) A cytoarchitectonic atlas of the spinal cord in the cat. J Comp Neurol 100:297–379
Rice ME, Nicholson C (1988) Behavior of extracellular K^+ and pH in skate (*Raja erinacea*) cerebellum. Brain Res 461:328–334
Robertson D, Johnstone BM (1979) Aberrant tonotopic organisation in the inner ear damaged by kanamycin. J Acoust Soc Am 66:466–469
Robinson, JD (1968) Regulating ion pumps to control cell volume. J Theor Biol 19:90–96
Roitbak AI (1988) Neuroglia: properties, functions and significance in nervous activity. Sov Sci Rev Physiol Gen Biol 2:355–402
Roos A, Boron WF (1981) Intracellular pH. Physiol Rev 61:296–434
Rosenthal M, Somjen G (1973) Spreading depression, sustained potential shifts and metabolic activity of cerebral cortex of cats. J Neurophysiol 36:739–749
Rosenthal M, LaManna J, Yamada S, Younts W, Somjen G (1979) Oxidative metabolism, extracellular potassium and sustained potential shifts in cat spinal cord in situ. Brain Res 162:113–127
Roussel G, Delaunoy JP, Musstaum JL, Mandel P (1979) Demonstration of a specific localization of carbonic anhydrase C in the glial cells of rat CNS by an immunohistochemical method. Brain Res 169:47–55
Russell IJ (1983) The origin of receptor potential in inner hair cells of the mammalian cochlea – evidence for Davis's theory. Nature 301:334–336
Russell JM, Boron WF, Brodwick MS (1983) Intracellular pH and Na fluxes in barnacle muscle with evidence for reversal of the ionic mechanism of intracellular pH regulation. J Gen Physiol 82:47–78
Salem RD, Hammerschlag R, Bracho H, Orkand RK (1975) Influence of potassium ions on accumulation and metabolism of $[^{14}C]$ glucose by glial cells. Brain Res 86:499–503

Satoh T, Watanabe K, Eguchi K (1979) Enhancement during REM sleep of extracellular potassium ion activity in the reticular formation. Brain Res 174:180–183

Scheibel ME, Scheibel AB (1968) Terminal axonal patterns in the cat spinal cord. II. The dorsal horn. Brain Res 9:32–58

Schlue WR, Deitmer JW (1988) Ionic mechanisms of intracellular pH regulation in the nervous system. In: Thomas RC (ed) Proton passage across membranes. Wiley, Chichester, pp 47–69

Schmidt RF (1971) Presynaptic inhibition in the vertebrate central nervous system. Ergeb Physiol 63:20–101

Schmitt FO, Samson FE Jr (1969) Brain cell microenvironment. Neurosci Res Program Bull 7:277–417

Schousboe A, Booher J, Hertz L (1970) Content of ATP in cultivated neurons and astrocytes exposed to balanced and potassium-rich media. J Neurochem 17:1501

Schulthess P, Shijo Y, Pham HV, Pratsch E, Ammann D, Simon W (1981) A hydrogen ion-selective liquid-membrane electrode based on tri-*n*-dodecylamine as neutral carrier. Anal Chim Acta 131:111–116

Schwartzkroin PA, Prince DA (1978) Cellular and field potential properties of epileptogenic hippocampal slices. Brain Res 147:117–130

Sears TA (ed) (1982) Neuronal-glial cell interrelationships. Springer, Berlin Heidelberg New York

Segal M, Gutnick MJ (1980) Effects of serotonin on extracellular potassium concentration in the rat hippocampal slice. Brain Res 195:389–401

Sellick PM, Bock GR (1974) Evidence for an electrogenic potassium pump as the origin of the positive component of the endocochlear potential. Pflügers Arch 352:351–361

Shanes AM (1954) Effect of temperature on potassium liberation during nerve activity. Am J Physiol 177:377–382

Sick JT, Kreisman NR (1981) Potassium ion homeostasis in amphibian brain: contribution of active transport and oxidative metabolism. J Neurophysiol 45:998–1012

Siemkowicz E, Hansen AJ (1981) Brain extracellular ion composition and EEG activity following 10 minutes ischemia in normo- and hyperglycemic rats. Stroke 12:236–240

Siesjö BK (1981) Cell damage in the brain: a speculative synthesis J Cereb Blood Flow Metab 1:155–185

Siesjö BK (1984) Cerebral circulation and metabolism. J Neurosurg 60:883–908

Siesjö BK, von Harewehr R, Nergelius G, Nevander G, Ingwar M (1985) Extra- and intracellular pH in the brain during seizures and in the recovery period following the arrest of seizure activity. J Cereb Blood Flow Metab 5:47–57

Silver IA (1978) Cellular microenvironment in relation to local blood flow. Ciba Found Symp 56:49–67

Simon W, Pretsch E, Morf WE, Ammann D, Oesch U, Dinten O (1984) Design and application of neutral carrier-based ion-selective electrodes. Analyst 109:207–209

Sims TJ, Waxman SG, Black JA, Gilmore SA (1985) perinodal astrocytic processes at nodes of Ranvier in developing normal and glial cell deficient rat spinal cord. Brain Res 337:321–333

Singer W, Lux HD (1975) Extracellular potassium gradients and visual receptive fields in the cat striate cortex. Brain Res 96:378–383

Škvařil J, Radil-Weiss T, Bohdanecký Z, Syka J (1971) Spontaneous discharge patterns of mesencephalic neurons. Interval histogram and mean interval relationship. Kybernetik 9:11–15

Šramka M, Brožek G, Bureš J, Nádvorník P (1977/1978) Functional ablation by spreading depression: possible use in human stereotactic neurosurgery. Appl Neurophysiol 40:48–61

Smith CA, Lowry OH, Wu Mei-Ling (1954) The electrolytes of the labarynthine fluids. Laryngoscope 64:141–153

Somjen GG (1970) Evoked sustained focal potentials and membrane potential of neurons and of unresponsive cells of the spinal cord. J Neurophysiol 33:562–582
Somjen GG (1973) Electrogenesis of sustained potentials. Prog Neurobiol 1:201–237
Somjen GG (1975) Electrophysiology of neuroglia. Annu Rev Physiol 37:163–190
Somjen GG (1979) Extracellular potassium in the mammalian central nervous system. Annu Rev Physiol 41:159–177
Somjen GG (1984) Acidification of interstitial fluid in hippocampal formation caused by seizures and by spreading depression. Brain Res 311:186–188
Somjen G, Rosenthal M, Cordingley G, LaManna JC, Lothman E (1976) Potassium, neuroglia and oxidative metabolism in central gray matter. Fed Proc 35:1266–1271
Sonnhof U (1987) Single-voltage-dependent K^+ and Cl^- channels in cultured rat astrocytes. Can J Physiol Pharmacol 65:1043–1050
Sonnhof U, Bührle CP (1980) On the postsynaptic action of glutamate in frog spinal motoneurons. Pflugers Arch 388:101–109
Sonnhof U, Schachner M (1986) Single voltage-dependent K^+-channels in cultured astrocytes. Neurosci Lett 64:241–246
Sotnikov OS, Roitbak AI (1980) Fast functional myelinization stimulated by potassium ions. Dokl Akad Nauk SSSR 255:986–988
Spira ME, Yarom Y, Zeldes D (1984) Neuronal interactions mediated by neurally evoked changes in the extracellular potassium concentration. J Exp Biol 112:179–197
Spray DC, Harris AL, Bennett MVL (1981) Gap junctional conductance is a simple and sensitive function of intracellular pH. Science 211:712–715
Steinberg RH, Oakley B, Niemeyer G (1980) Light-evoked changes in $[K^+]_o$ in retina of intact cat eye. J Neurophysiol 44:897–921
Steiner RA, Oehme M, Ammann D Simon W (1979) Neutral carrier sodium ion-selective microelectrode for intracellular studies. Anal Chem 51:351–353
Stewart PA (1978) Independent and dependent variables of acid-base control. Respir Physiol 33:9–26
Stewart PA (1981) How to understand acid base. Elsevier, New York
Stöckle H, Bruggencate ten G (1978) Climbing fiber-mediated rhythmic modulations of potassium and calcium in cat cerebellar cortex. Exp Neurol 61:226–230
Stöckle H, Bruggencate ten G (1980) Fluctuation of extracellular potassium and calcium in the cerebellar cortex related to climbing fiber activity. Neuroscience 5:893–901
Sugaya E, Takato M, Noda Y (1975) Neuronal and glial activity during spreading depression in cerebral cortex of rat. J Neurophysiol 38:822–841
Sugimoto M, Kuraishi Y, Satoh M, Takagi H (1986) Involvement of medullary opioid - peptidergic and spinal noradrenergic systems in the regulation of formalin-induced persistent pain. Neuropharmacology 25:481–485
Sussdorf WS, Campenot RB (1986) Influence of the extracellular potassium environment on neurite growth in sensory neurons, spinal cord neurons and sympathetic neurons. Dev Brain Res 25:43–52
Suzuki T, Oka J-I, Nagano N, Fukuda H (1986) Enkephalin but not morphine modulates the motor activity in the frog spinal cord in vitro. Comp Biochem Physiol 83C:245–251
Svoboda J, Syková E (1989) Activity-related extracellular K^+ and volume changes in spinal cord of the rat. Acta Physiol Scand 136 [Suppl 582]:86
Svoboda J, Syková E (1991) Extracellular space volume changes in the rat spinal cord produced by nerve stimulation and peripheral injury. Brain Res 560:216–224
Svoboda J, Motin V, Hájek I, Syková E (1988) Increase in extracellular potassium level in rat spinal dorsal horn induced by noxious stimulation and peripheral injury. Brain Res 458:97–105
Svoboda J, Syková E, Motin V (1989) The extracellular space volume changes in the rat spinal cord evoked by electrical stimulation and peripheral injury, Physiol Bohemoslov 38:376

Svoboda J, Syková E, Drozen D (1990) Stimulation-related changes in extracellular volume in the rat spinal cord, Physiol Bohemoslov 39:539
Syka J, Popelář J, Radil-Weiss T (1975) Influence of increasing doses of pentobarbital on the mesencephalic reticular formation in rats. Spontaneous firing of neuronal pairs and activity evoked by polarization. Brain Res 88:263–279
Syková E (1976) Změny koncentrace K^+ v extracelulárním prostoru míchy při vzruchové aktivitě (Activity-related extracellular K^+ changes in spinal cord) Ph D dissertation
Syková E (1978) GABA-induced changes of extracellular K^+-activity in the frog spinal cord. Physiol Bohemoslov 27:189–192
Syková E (1981) K^+ changes in the extracellular space of the spinal cord and their physiological role. J Exp Biol 95:93–109
Syková E (1983) Extracellular K^+ accumulation in the central nervous system. Prog Biophys Mol Biol 42:135–189
Syková E (1986) Activity-related fluctuations in extracellular ion concentrations in the central nervous system, NIPS 1:57–61
Syková E (1987) Modulation of spinal cord transmission by changes in extracellular K^+ activity and extracellular volume. Can J Physiol Pharmacol 65:1058–1066
Syková E (1989) Activity-related extracellular pH transients in spinal cord. Verh Dtsch Zool Ges 82:153–163
Syková E, Orkand RK (1980) Extracellular potassium accumulation and transmission in frog spinal cord. Neuroscience 5:1421–1428
Syková E, Svoboda J (1990) Extracellular alkaline-acid-alkaline transients in the rat spinal cord evoked by peripheral stimulation. Brain Res 512:181–189
Syková E, Vyklický L (1977a) Changes of extracellular potassium activity in isolated spinal cord of frog under high Mg^{2+} concentration. Neurosci Lett 4:161–165
Syková E, Vyklický L (1977b) GABA effect on extracellular K^+ activity in the frog spinal cord. Physiol Bohemoslov 26:469
Syková E, Vyklický L (1978) Effects of picrotoxin on potassium accumulation and dorsal root potentials in the frog spinal cord. Neuroscience 3:1061–1067
Syková E, Vyklický L (1979) Isolated spinal cord of the frog – an in vitro model for the study of neuronal mechanisms of pain. Physiol Bohemoslov 28:227–230
Sýkorová (Syková) E, Škvařil J, Radil-Weiss T (1968) Interrelationship between spontaneous impulses activity of reticular neurones and cyclic cardiac activity. Physiol Bohemoslov 17:490
Syková E, Rothenberg S, Krekule I (1974) Changes of extracellular potassium concentration during spontaneous activity in the mesencephalic reticular formation of the rat. Brain Res 79:333–337
Syková E, Shirayev B, Kříž N Vyklický (1976) Accumulation of extracellular potassium in the spinal cord of frog. Brain Res 106:413–417
Syková E, Czéh G Kříž N (1980) Potassium accumulation in the frog spinal cord induced by nociceptive stimulation of the skin. Neurosci Lett 17:253–258
Syková E, Hník P, Vyklický L (eds) (1981) Ion-selective microelectrodes and their use in excitable tissues. Plenum, New York
Syková E, Kříž N, Preis P (1983) Elevated extracellular potassium concentration in unstimulated spinal dorsal horns of frogs. Neurosci Lett 43:293–298
Syková E, Hájek I, Chvátal A, Kříž N, Diatchkova GI (1985a) Changes in extracellular potassium accumulation produced by opioids and naloxone in frog spinal cord: relation to changes of Na-K pump activity. Neurosci Lett 59:285–290
Syková E, Kříž N, Hájek I (1985b) Role of endogenous opiates and extracellular K^+ accumulation in the inhibition of frog spinal reflexes by electrical skin stimulation. Physiol Bohemoslov 34:548–561
Syková E, Kříž N, Chvátal A, Motin V, Hájek I (1987a) Activity-related transient changes in the sizes of the extracellular space in the frog spinal cord. Physiol Bohemoslov 36:379

Syková E, Syka J, Patuzzi R, Johnstone BM (1987b) Potassium concentration changes in the organ of Corti evoked by acoustic stimulation. Physiol Bohemoslov 36: 559

Syková E, Orkand RK, Chvátal A, Hájek I, Kříž N (1988a) Effects of carbon dioxide on extracellular potassium accumulation and volume in isolated frog spinal cord. Pflügers Arch 412:183–187

Syková E, Svoboda J, Chvátal A, Jendelová P (1988b) Extracellular pH and stimulated neurons. In: Thomas RC (ed) Proton passage across cell membranes. Wiley, Chichester, pp 220–232

Syková E, Jendelová P, Svoboda J, Sedman G, Ng KT (1990) Activity-related rise in extracellular potassium concentration in the brain of 1–3-day-old chicks. Brain Res Bull 24:569–575

Sypert GW, Ward AA Jr (1974) Changes in extracellular potassium activity during neocortical propagated seizures. Exp Neurol 45:19–41

Székely G (1976) The morphology of motoneurons and dorsal root fibers in the frog's spinal cord. Brain Res 103:275–290

Takahashi T, Tsuruhara H (1987) Slow depolarizing potentials recorded from glial cells in the rat superficial dorsal horn. J Physiol (Lond) 388:597–610

Takeuchi A, Takeuchi N (1962) Electrical changes in pre- and post-synaptic axons of the giant synapse of Loligo. J Gen Physiol 45:1181–1193

Tang C-M, Cohen MV, Orkand RK (1980) Electrogenic pumps in axons and neuroglia and extracellular potassium homeostasis. Brain Res 194:283–286

Tang C-M, Orkand PM, Orkand RK (1985) Coupling and uncoupling of amphibian neuroglia. Neurosci Lett 54:237–242

Tas PWL, Massa PT, Koschel K (1986) Preliminary characterization of an Na^+ K^+ Cl^- contransport activity in cultured human astrocytes. Neurosci Lett 70:369–373

Tasaki I, Byrne PM (1983) Swelling of frog dorsal root ganglion and spinal cord produced by afferent volley of impulses. Brain Res 272:360–363

Tasaki I, Iwasa K (1982) Further studies of rapid mechanical changes in squid giant axon associated with action potential production. Jpn J Physiol 32:505–518

Thomas RC (1976) Ionic mechanism of the H^+ pumps in a snail neurone. Nature 262:54–55

Thomas RC (1977) The role of bicarbonate, chloride and sodium ions in the regulation of intracellular pH in snail neurones. J Physiol (Lond) 273:317–338

Thomas RC (1978) Ion-sensitive intracellular microelectrodes. Academia London

Thomas RC (1984a) Experimental displacement of intracellular pH and the mechanism of its subsequent recovery. J Physiol (Lond) 354:3P–22P

Thomas RC (1984b) Electrogenic sodium pumps current associated with recovery from intracellular acidification of snail neurons. In: Mordecai P, Blaustein P, Lieberman M (eds) Electrogenic transport: fundamental principles and physiological implications. Raven, New York, pp 353–363

Thomas RC (1988a) Changes in the surface pH of voltage-clamped snail neurones apparently caused by H^+ fluxes through a channel. J Physiol (Lond) 398:313–324

Thomas RC (1988b) Proton passage across cell membranes. Wiley, Chichester

Thomas RC (1988c) Proton channels in snail neurons studied with surface pH glass microelectrodes. In: Thomas RC (ed) Proton passage across cell membranes. Wiley, Chichester, pp 168–183

Tomita T, Yanagida T (1981) Origins of the ERG waves. Vision Res 21:1703–1707

Treherne JE (ed) (1981) Glial-neurone interactions. J Exp Biol 95:1–240

Trubatch J, Loud AU, Van Harreveld A (1977) Quantitatave stereological evaluation of KCl-induced ultrastructural changes in frog brain. Neuroscience 2:963–974

Tsacopoulos M, Levy S (1976) Intraretinal acid-base studies using pH glass microelectrodes: effect of respiratory and metabolic acidosis and alkalosis on inner-retinal pH. Exp Eye Res 23:495–504

Tuckwell HA, Miura RM (1978) A mathematical model for spreading cortical depression. Biophys J 23:257–276

Turin L, Warner AE (1980) Intracellular pH in early Xenopus embryons: its effects on current flow between blastomeres. J Physiol (Lond) 300:489–504
Ujec E (1988) Differential DC amplifier for recording small and fast concentration changes with ion-selective microelectrodes. Physiol Bohemoslov 37:87–90
Urbán L, Aitken PG, Somjen GG (1985) Interstitial potassium concentration, slow depolarization and focal potential responses in the dorsal horn of the rat spinal slice. Brain Res 331:168–171
Urbanics R, Leniger-Follert E, Lübbers DW (1978) Time course of changes of extracellular H^+ and K^+ activities during and after direct electrical stimulation of the brain cortex. Pflügers Arch 378:47–53
Van Harreveld A (1959) Component of brain extracts causing spreading depression of cerebral cortical activity and crustacean muscle. J Neurochem 3:300–315
Van Harreveld A (1966) Brain tissue electrolytes. In: Bittar EE (ed) Molecular biology and medicine series. Butterworths, London
Van Harreveld A (1972) The extracellular space in the vertebrate central nervous system. In: Bourne GH (ed) The structure and function of nervous tissue. Academic, New York, pp 449–511
Van Harreveld A, Fifková E (1974) Involvement of glutamate in memory formation. Brain Res B1:455–467
Van Harreveld A, Fifková E (1975) Swelling of dendritic spines in the fascia dentata after stimulation of the perforant fibres as a mechanism of post-tetanic potentiation. Exp Neurol 49:736–749
Van Harreveld A, Khattab FI (1967) Changes in cortical extracellular space during spreading depression investigated in the electron microscope. J Neurophysiol 30:911–929
Van Harreveld A, Trubatch J (1974) Conditions affecting the extracellular space in the frog's forebrain. Anat Res 178:587–597
Vaughan-Jones RD, Lederer WJ, Eisner DA (1983) Ca^{2+} ions can affect intracellular pH in mammalian cardiac muscle. Nature 303:522–524
Verzeano H, Calma I (1954) Unit activity in spindle bursts. J Neurophysiol 17:417–428
Vodovnik L (1981) Indirect spinal cord stimulation. Appl Neurophysiol 44:97–113
Vyklický L, Syková E (1980) Effects of increased K^+ activity on the flexor reflex in the isolated spinal cord of the frog. Neurosci Lett 19:203–207
Vyklický L, Syková E, Kříž N, Ujec E (1972) Post-stimulation changes of extracellular potassium concentration in the spinal cord of the rat. Brain Res 45:608–611
Vyklický L, Syková E, Kříž N (1975) Slow potentials induced by changes of extracellular potassium in the spinal cord of the cat. Brain Res 87:77–80
Vyklický L, Syková E, Mellerová B (1976) Depolarization of primary afferents in the frog spinal cord under high Mg^{2+} concentrations. Brain Res 117:153–156
Vyklický L, Vyskočil F, Kolaj M, Jastreboff P (1982) Primary afferent depolarization and changes in extracellular potassium concentration induced by L-glutamate and L-proline in the isolated spinal cord of the frog. Neurosci Lett 32:159–164
Vyskočil F, Kříž N (1972) Modification of single- and double-barrel potassium specific microelectrodes for physiological experiments. Pflügers Arch Gen Physiol 337:265–276
Vyskočil F, Kříž N, Bureš J (1972) Potassium-selective microelectrodes used for measuring the extracellular brain potassium during spreading depression and anoxic depolarization in rats. Brain Res 39:255–259
Walker JL Jr (1971) Ion specific liquid ion exchanger microelectrodes. Anal Chem 43:89A–93A
Wall PD (1958) Excitability changes in afferent fibre terminations and their relation to slow potentials. J Physiol (Lond) 142:1–21
Wall PD (1962) The origin of a spinal cord slow potential. J Physiol (Lond) 164:508–526
Wall PD (1964) Presynaptic control of impulses at the first central synapse in the cutaneous pathway. In: Prog Brain Res 12:92–118

Wall PD (1978) The gate control theory of pain mechanisms. A re-examination and re-statement. Brain 101:1–18

Walton DK, Chesler M (1988) Activity-related extracellular potassium transients in the neonatal rat spinal cord: an in vitro study. Neuroscience 25:983–995

Walz W (1989) Role of glial cells in the regulation of the brain ion microenvironment Prog Neurobiol 33:309–333

Walz W, Hertz L (1982) Ouabain-sensitive and ouabain-resistant net uptake of potassium into astrocytes and neurons in primary cultures. J Neurochem 39:70–77

Walz W, Hertz L (1983) Intracellular ion changes of astrocytes in response to extracellular potassium. J Neurosci Res 10:411–423

Walz W, Hertz L (1984) Intense furosemide-sensitive potassium accumulation in astrocytes in the presence of pathologically high extracellular potassium levels. J Cereb Blood Flow Metab 4:301–304

Walz W, Hinks EC (1986) A transmembrane sodium cycle in astrocytes. Brain Res 368:226–232

Walz W, Hinks EC (1987) Extracellular hydrogen ions influence channel-mediated and carrier-mediated K^+ fluxes in cultured mouse astrocytes. Neuroscience 20:341–346

Walz W, Mukerji S (1988a) KCl movements during potassium-induced cytotoxic swelling of cultured astrocytes. Exp Neurol 99:17–29

Walz W, Mukerji S (1988b) Lactate production and release in cultured astrocytes. Neurosci Lett 86:296–300

Wang W, Dietl P, Oberleithner H (1987) Evidence for Na^+ dependent rheogenic HCO_3^- transport in fused cells of frog distal tubules. Pflügers Arch 408:291–299

Waniewski RA, Martin DL (1983) Selective inhibition of glial versus neuronal uptake of L-glutamic acid by SITS. Brain Res 268:390–394

Weight FF, Erulkar SD (1976) Modulation of synaptic transmitter release by repetitive postsynaptic action potentials. Science 193:1023–1025

Whistler JW, Johnson D (1978) Epileptogenesis: a model for the involvement of slow membrane events and extracellular potassium. J Theor Biol 75:271–288

Winkler BS (1986) Buffer dependence of retinal glycolysis and ERG potentials. Exp Eye Res 42:585–593

Wolpaw EW, Martin DL (1984) Cl^- transport in a glioma cell line: evidence for two transport mechanisms. Brain Res 297:317–327

Wong RKS, Prince DA, Basbaum AI (1979) Intradendritic recordings from hippocampal neurons. Proc Natl Acad Sci USA 76:986–990

Woolf CJ (1979) Transcutaneous electrical nerve stimulation and the reaction to experimental pain in human subjects. Pain 7:115–127

Woolf CJ (1983) Evidence for a central component of post-injury pain hypersensitivity. Nature 306:686–688

Wright EM (1972) Mechanisms of ion transport across the chorioid plexus. J Physiol (Lond) 226:545–571

Wuhrmann P, Ineichen H, Riesen-Willi U, Letzi M (1979) Change in nuclear potassium electrochemical activity and puffing of potassium-sensitive salivary chromosome regions during *Chironomus* development. Proc Natl Acad Sci USA 76:806–808

Yaksh TL (1981) Spinal opiate analgesia: characteristics and principles of action. Pain 11:293–346

Yamate CL, Ransom BR (1985) Effects on altered gliogenesis on activity-dependent K^+ accumulation in the developing rat optic nerve. Devl. Brain Res 21:167–173

Zenner HP, Zimmermann N, Schmitt U (1985) Reversible contraction of isolated mammalian cochlear hair cells. Hear Res 18:127–134

Zuckerman R (1973) Ionic analysis of photoreceptor membrane currents. J Physiol (Lond) 235:333–354

Zuckermann EC, Glaser GH (1968) Hippocampal epileptic activity induced by localized ventricular perfusion with high-potassium cerebrospinal fluid. Exp Neurol 20:87–110

Subject Index

MIX
Papier aus verantwortungsvollen Quellen
Paper from responsible sources
FSC® C105338

If you have any concerns about our products,
you can contact us on
ProductSafety@springernature.com

In case Publisher is established outside the EU,
the EU authorized representative is:
Springer Nature Customer Service Center GmbH
Europaplatz 3, 69115 Heidelberg, Germany

Printed by Libri Plureos GmbH
in Hamburg, Germany